HEARTS AND MINDS

The story of the women of Nottinghamshire in the Miners' Strike, 1984-1985

Joan Witham

Canary Press

Hearts and Minds

ISBN 0 9509967 7 7

Published by Canary Press
First Edition September 1986

Typeset by Grassroots, London NW6
01-328 0318
Printed in England by Biddles, Guildford

Cover design by Graphics International, London N1
01-833 2988

Cover photograph: Cath Black, Ollerton Women's Support Group,
Striking Canteen Worker from Bevercotes
Back: Picket at Bevercotes

DEDICATION

It is only fitting that this history of the Nottinghamshire women in the 1984-85 miners' strike be dedicated to Liz Hollis, an active member of the Nottinghamshire Women's Support Group throughout the dispute. It was hoped, right up until her tragic death on February 16th, 1986, that her own written contribution would have been available.

There is no one who would deny that the achievements of the Nottinghamshire women were in no small part due to the influence of Liz. She gave more than seemed humanly possible during the epic struggle, working 16 to 18 hours a day, 7 days a week, for over a year. She undertook every task; from organising the office to delivering logs; from arranging holidays for the strikers and their families to speaking at meetings; from preparing and printing campaign materials to picketing; from meeting visitors to baby-sitting.

Invariably she would arrive for meetings half way through the first item, an infectious grin on her face as she said, "Sorry I'm late". No matter how frustrated one might feel before she arrived, her presence melted all irritation away and the meeting was transformed by her enthusiasm and commitment. Despite her almost habitual lateness she would always turn up and everything would get done in the end.

Her small physique belied her stature; she was a leader, a teacher and a friend who gave confidence to us all, loved and respected by everyone who knew her. It was a privilege and an honour to have known her and worked with her. We will cherish her memory always.

Ida Hackett.

Ida Hackett
(for Nottinghamshire Women's Support Group) April 1986

ACKNOWLEDGEMENTS

I should, really, like to thank by name all the women in the Nottinghamshire Miners' Support Groups, for giving me not only the information for the book, but also their friendship and acceptance during the strike. Since it is impossible to name them all I can only offer a general debt of gratitude.

I am grateful to my husband, Jef, who worked with me during the strike and who encouraged me to write the book, doing all the household chores instead of only his half share whilst I researched and wrote up my material.

The work done by Babs Segal in editing the manuscript and arranging typesetter and printer for me is much appreciated. My thanks are also due to Andrew Burgin who was always encouraging and helpful, and the workers at Housmans Bookshop for all their help.

To those generous individuals and trade unions who gave me donations to finance production of the book, I give my thanks: Mr. and Mrs. Alsop, Joan Maynard MP, Gaving Strang MP, Campaign Group of MP's, Notts Miners Forum, Nottinghamshire NUM, ASLEF, TASS, The Tobacco Worker's Union, NUPE and Worksop and District Trades Council. Other donations which came too late for inclusion here are no less appreciated.

FOREWORD

The 1984/85 miners' strike was one of the most bitter struggles in the history of the British working class. Now and in the future, many authors have and will attempt to record and analyse the political and social consequences of that dispute.

This book is part of that analysis, but concentrates on one particular aspect of the dispute—the emergence of the women, their role and their politicisation—in the Nottinghamshire coalfield, where the striking miners were in a minority and faced particular difficulties.

It is not the author's intention to diminish the role played by women in other Areas, but to catalogue the Nottinghamshire women's movement as she experienced it, both as an observer and a participant. Without the women, who encouraged and supported their menfolk through the very difficult twelve months, many more of the strikers in Nottinghamshire would undoubtedly have given in.

The organisation of the food kitchens, with barely any financial support from the Nottinghamshire Area NUM was, in itself, a magnificent achievement; and the bonds of solidarity made with other women's groups, both nationally and internationally, are cemented forever.

No Nottinghamshire striking miner and his family will ever forget the wonderful Christmas of 1984, when thanks largely to the women's groups, they all enjoyed festivities and presents for the children which many say were the best they ever had. The working miner may not have lost his Christmas bonus from the National Coal Board, but he lost something of far greater value, the experience of comradeship and working-class solidarity.

The Nottinghamshire women, like their counterparts in other Areas, joined their menfolk on the picket lines. Some were arrested, many were verbally and physically abused and intimidated. None gave up because of this, but stood by their principles, firm in the knowledge that they were right. As the dispute progressed, their determination to continue the fight for the jobs of their husbands, sons and grandsons became stronger. They are still determined to remain organised so as to resist the constant erosion of their rights and other attacks on working-class people.

I personally am particularly grateful to all the women who helped, supported and encouraged me in the difficult time I endured when General Secretary at Berry Hill. I would like to believe that together we were able to help our miners' families as far as was humanly possible, both financially and morally, throughout the strike, despite the intereference and hostility which emanated from the majority at Berry Hill.

One book cannot adequately relate all that the women achieved in Nottinghamshire, but it is essential that something is written down, in order that we can look back, contemplate, and give our heartfelt thanks.

I hope that this book can do that and also remind the reader that this was part of just one battle in a long war. The women were a regiment in the forefront of that battle.

THE FIGHT GOES ON

Henry Richardson
General Secretary—Nottinghamshire Area NUM
April 1986

PREFACE

At no time in my life did I think that I would write a book. I didn't want to write *this* book, but I thought it important that *someone* should record the role of the women of Nottinghamshire in the strike. I wanted to ensure that this piece of history was not lost, and in particular that it was not lost to other women.

When the miners came out on strike in 1984 I did not for one moment anticipate that I would become so involved in the dispute. But the more I did, the more I learned, and the more I wanted to do and to learn. It was a snowball effect, and I ended up giving most of the year to the strike.

Having retired in 1981, I was going to tuck myself away and have a quiet, lazy life in a new area, Nottinghamshire, while continuing my involvement in the Labour Party and the peace movement. Though I have been keenly interested in politics since my schooldays, I had never been particularly active until my retirement.

I grew up in the village of Tupton near Chesterfield. Some of the village men worked in nearby Grassmoor pit and there was a small drift mine in the village. As I child I had watched the men and the tiny pony disappear into the gaping black hole, terrified that they would be swallowed up, never to emerge again. It was always a relief to see the little pony trotting round our streets pulling a cart full of outcrop coal for delivery. On the way home from school I watched the off-shift miners squatting by the wall, fascinated by their black-rimmed eyes and the bits of coal embedded in their faces. I heard them cough and saw them spit, only realising much later that they were coughing out the coal dust which was rotting their lungs.

My father was not a miner and we were not part of the mining community in the village, but my mother's father had been killed by a roof fall at Park House No. 2 pit in 1922. He died at the age of forty eight, leaving my grandmother to bring up seven children alone.

Years later, as head teacher of a small school in the pit village of Pinxton, Derbyshire, I taught the children of miners, but it was not until the strike that I really began to know mining people, to learn about their industry, their patterns of life and their tough, independent spirit. At the start of the dispute I had no special allegiance to the miners; but as a result of my experience I have a special affinity with them now.

In my hometown of Newark, Edna and Richard Colston, with Mary and Merf Roberts, decided early in the strike that we ought to raise some money for the Nottinghamshire Miners' Forum. This first social event raised about £150 and brought together a group of people who wanted to do more. It was very difficult to find out exactly what was happening in those early days in Nottinghamshire, but we formed a support group and made contact with about ten local miners on strike.

I became secretary of the group and as I seemed to have more time than most people I started visiting Blidworth, Rainworth and Bilsthorpe, the three pits where our Newark miners were employed. Though Blidworth support group asked us to give all our help to Rainworth and Bilsthorpe, we continued to visit the 'Glasshouse' in Blidworth every week. I also became a delegate to the central Nottinghamshire Women's Support Group.

My entry into the women's groups was slow and tentative. For several months I felt very much on the edge of things. A natural talker, I had to try to keep quiet and learn from other women as much as I could about their attitudes and their plans. Doreen Humber told me later that they all wondered who this 'Tory-looking' woman was, and what I was doing there. The men were equally suspicious. According to John Eason, one of the Newark miners, they regarded me as 'a bit of a feminist'. They thought I was encouraging the women to become too bossy and Pip Humber jokingly threatened to put a picket on the village after the strike in order to keep me out! Later he realised that Doreen and the other Blidworth women didn't need anyone to push them along, they were already well in control, developing their confidence at a tremendous pace, and were unstoppable. After other 'different' women, from London and Cambridge, had visited the group, and after I had turned up on the picket line and shown my commitment in other ways, I was accepted, though the good-humoured teasing and chaffing continued. When the men began to swear in front of me without apologising I knew that I was finally a part of the group!

As I became more confident, and as we began to get to know each other, I was able to offer my own ideas and suggestions. At first I found the direct, blunt way in which many of the women put things a little uncomfortable. In discussions with my colleagues in school and college I had grown used to depersonalising issues and avoiding confrontation. I sometimes watched the 'hammer and tongs' battles and thought how damaging they were. I was wrong. I learnt that it was this very directness and honesty which got things done so quickly and economically. But I never could do things this way myself, finding it impossible to break out of the pattern I have acquired over so many years of trying to be tactful and impersonal.

One of the strongest emotions I felt during my year of involvement was that of anger. I think it was this anger which gave me such a strong bond with the miners and their families. Their hatred of the scabs, the Coal Board, and later the police, got into my bloodstream too.

I have always believed that as workers our labour is of value in two ways; firstly for the goods we produce or the services we give, and secondly for the personal effort, time and sacrifice involved in the work. The personal contribution of miners and other workers in uncomfortable and dangerous industries is a very substantial one. They risk their lives and endanger their health on every shift. It seemed to me a terrible outrage that the livelihood of workers such as these should be made the subject of a vicious government attack. The miners, along with their wives and families, were being made to suffer for being a strong and militant workforce, in the vanguard of the Trade Union Movement in this country. They were to be disciplined and humiliated, systematically and ruthlessly, as an example to workers in other industries. Although I had been a union member all my working life, I had only been called out on strike for one day in thirty years. I had never known the pride of being in a strong union with long-standing traditions and practices. I had never been called upon to make great sacrifice for my union and on behalf of my colleagues and I felt the lack of that experience of solidarity. But I wanted to be part of the movement which tried to act as a bulwark against the onslaught that the miners were now experiencing.

After the strike ended, I began to talk to Jackie Naylor, Ida Hackett and Pam Oldfield about the importance of collecting material from all the groups in the county and putting it together in a book. The politics of the strike have been well documented elsewhere, but scant mention has been made of the men and women of Nottinghamshire who supported the strike and fought for their rights against the full weight of the state. I wanted to show how, despite their comparative isolation, the strikers and their wives worked with just as much loyalty, endurance and courage as their comrades in the 'solid' coalfields. The very fact that they were such a minority demands that they be given special recognition. This book is an attempt to set the record straight, at least on behalf of the women. The men's story should also be told, but that is for another author.

Based on observations, conversations and interviews with women from the 30 groups which formed the Nottinghamshire Women's Support Group, the book seeks to tell the story of their involvement in the year long dispute. I hope that it gives an accurate account of the work done and of the attitudes, anxieties, relationships and perspectives of the women involved. I also want it to be a tribute to them and to their children. They were so courageous, so enduring and so determined.

Joan Witham

CONTENTS

INTRODUCTION
The Strike in Nottinghamshire

Throughout the period of the strike and in the months immediately following the return to work, the general picture of the Nottinghamshire coalfield presented by the media was of a workforce which had turned its defiant back on the struggle of fellow miners and on the National Union of Mineworkers. In the aftermath of the strike, as the move by the rebel Nottinghamshire Area Council to set up a breakaway union gained momentum, the Nottinghamshire miners were portrayed as being almost totally united in their determination to seek 'democracy' in a union divorced from the NUM.

This blanket portrayal of Nottinghamshire miners as strike breakers, or 'men loyal to their industry and stalwarts in the struggle for democracy', as they were described by the government and the Coal Board, was, and is, bitterly resented by the men and women in this county who were on strike along with their comrades from the other British coalfields. It does grave injustice to the 10,000 men who joined the stoppage at the beginning and the 2,500 who remained loyal to the bitter end. Sadly, even strikers and their supporters from the other coalfields often included Nottinghamshire miners as a whole in their hatred of strike breakers, bitterly resenting the fact that the outcome of the strike was so largely under their control. Even the cheers for the Nottinghamshire contingents on national marches were often accompanied by a degree of surprise that any of them were actually on strike. It was only at a personal level that Nottinghamshire strikers were accepted by men and women from Yorkshire, South Wales, Kent, Durham and Scotland.

The isolation of the Nottinghamshire strikers was an added burden for them throughout the dispute. But as the drift back to work progressed they felt increasingly alone as they became subjected to growing abuse and hostility.

When it became clear that the dispute was likely to be much longer than many had originally thought and all had hoped, the strikers' hostility towards their colleagues who continued to work grew into a festering sore of bitter rancour. With the realisation that the large numbers of miners working in the county was a major contributing factor to this situation the anatagonism escalated still further. The working miners retaliated and daily encounters were laced with venomous abuse of each other and sometimes worse.

Traditionally close family ties broke down, causing great distress and bitter recriminations. There were instances of sons on strike living in the same house as fathers who carried on working. Family events such as weddings, christenings and funerals highlighted these differences. One woman recalls an incident at a wedding when children called a grandfather "scab" and refused to have anything to do with him. The stress of such broken relationships weighed particularly heavily on the women.

Communities were torn apart. Rifts occurred between men who had worked together as mates in the same pit team, or on the same shift; who had drunk together in the same pubs and clubs, watched or played football together, shared the same streets as children and adults; between women who had grown up as school friends, who had worked, shopped and had a social life together. School playgrounds became battlefields as children challenged and abused each other, defending their fathers for whichever side they had taken.

At every place it was 'us' and 'them'. Neighbours exchanged heated arguments. People lowered their eyes and averted their faces, cut each other dead in every public place. Abusive, malicious and often threatening letters were sent to men and women active in the strike. Everyone took sides and argued their case with abuse and sometimes with physical violence.

Outside the pit areas, former friends and colleagues were bitterly divided as were some Labour Party and council groups in Nottinghamshire. Then gradually, apart from an occasional flare-up, the protagonists settled down to a mood of cold hostility. The strikers, their wives, and those children who understood, longed to be in places like South Wales, Kent or Yorkshire, to feel, instead of enmity and bruised bitterness, the solidarity of the mass.

These feelings of isolation were particularly marked when, at the end, the 'solid' areas saw their miners back to work *en masse,* with banners proudly proclaiming refusal to concede defeat, with songs

Dear friends

Please accept this donation to your just course. The work you are doing in support of your menfolk is nothing short of magnificent.

I am filled with great pride to see working class people at last fighting back against this terrible anti working class government.

The butcher MGregor must be defeated at all costs. The alternative to this does not bare thinking about. You and your menfolk are indeed leaders in the fight back against this Thatcher dictatorship.

I only wish the strike was one hundred per cent. How ever the men that are working will have to face reallity one of these days. They must feel a great sense of shame and guilt when they look in the mirror.

I wish you all a great succeus in these hard times. With unity and spirit you will certainly win.

Yours sincerely

Guernsey, Channel Islands. May 1984

and with proud dignity. In Nottinghamshire most men returned quietly, in ones and twos, to individual interviews with the pit manager. Many were directed to the most menial, low-paid and tedious jobs, punishment and intended degradation for their year of loyalty, hardship and principled endurance, or 'insubordination, defiance and aggression', depending on who was interpreting the behaviour.

The media almost totally ignored the return of the 2,500 miners in Nottinghamshire, continuing the lie that there were no strikers left in the county. The women witnessed the return with feelings of anguish, rage and biterness. They wept as they saw their proud heroes demeaned and threatened, at the same time applauding their continued defiance and determination, which they themselves shared. Together their conviction of being right all along was confirmed and strengthened as Nottingham became the focal point for anti-trade union activity, a break-away union and a hate campaign against the national leaders of the NUM.

WOMEN IN NOTTINGHAMSHIRE

Amongst the Nottinghamshire women who supported the strike the degree of involvement varied widely. The activists numbered only about two to three hundred, which meant, inevitably, that a few women did a great deal of work. Every time I called in at Bilsthorpe, for instance, there would be Janet Bone, stirring huge pans of stew or soup, her face reddened with the heat and steam of the tiny kitchen. Day in, day out, for the whole year, she and Joan Preston, Liz Dulley, Janice Shaw and Kay Stoneman cooked and served meals. The strain of doing all that was involved in providing this food, in addition to caring for their own homes and children, is hard to imagine. This picture was repeated throughout the support groups. In so many cases, there was absolutely no respite from the continual need to raise money, shop, budget and prepare food, all on a massive scale.

Some women, though supportive of the strike, were reluctant to be involved; others, like Esther O'Regan from Ollerton, were too shy. She stayed at home with her baby son throughout the dispute, longing to be part of the support group, but unable to pluck up enough courage to join the other women, who seemed to be so much more capable and confident than she was. Many women whose husbands were fully occupied in picketing or other strike activities had sole responsibility for small children. Some already had full or part-time jobs, or found work during the strike to help out with family income. Others were caring for aged or sick relatives. And some had husbands who were absolutely opposed to women participating in the strike.

The activists speak sadly of the many women who did not get involved: "We should have done more to find these women, we should have gone out to them; regardless of their problems we could have found something for them to do, even if it was very small," said Brenda Greenwood: "Our own lives were so enriched by the strike that we are sorry that many women missed out. Next time we should make it important to get more women involved."

Of course, those women who worked so hard for the strike had personal commitments too. They also had jobs, children, relatives to care for. Somehow, they managed to cope with both, drawing on strength that they did not know they had, trying to balance one commitment against another and to do justice to all of them.

Liz Gordon from Rainworth daily lugged her baby in a pram up the fifteen or so stairs to the cramped kitchen where she worked. Lyn Telford struggled up the same set of stairs, at first heavily pregnant, then complete with carry cot, scarcely taking time to recover from the birth before she was back again. All the women worried about neglecting their children. In particular, women who travelled outside Nottinghamshire fundraising sometimes had terrible feelings of guilt. While they were away from home, sometimes for periods of up to a week, the men had to take over the domestic chores. Some were willing, perhaps already used to sharing the care of children and of the home. Others were reluctant and still others really resentful. But for many men, there was no choice; the women said they were going and they went. Nevertheless they worried about both husbands and children left at home. Ann Jeffries, Glenys Pringle and Chris Gilbert went to London to speak at fire stations:

> Our little ones didn't understand what was happening. How could we explain why we had to go away? How could they realise that we couldn't always be there when they got in from school, or to tuck them up in bed at night? The dads were marvellous, but the kids missed us. We'd always been there before the strike. And we missed the kids. Before a meeting we'd say to each other, "Hope the kids are alright". We used to feel fair sad about them.

Lynne Dennet, one of the most active fundraisers, had a very young baby and five year old twins. She travelled to Holland with Debbie Bell and Alison Lovely to raise money:

> I think that towards the end all the kids of women who spent a lot of time away from home suffered, from the oldest to the youngest. It was such a change and so sudden. They may understand in time to come and we're making up for it now.

For some, the worry and resentment they felt about the effect of the strike on the children was compounded by what their youngsters encountered outside the home. Some faced abuse from the children of strike-breakers or from strike-breakers themselves. Others were harassed and questioned by the police. There were incidents, fortunately not common, of strikers' children being forced to stand separately from all the others to get their free meals. Children in some schools were denied outings or other treats because there was no money to pay for them. Joan Preston wrote in some notes for an article about her community on strike in Bilsthorpe:

> I would like you to give a thought for the children of the striker. So often they hear,
> "Sorry, no pocket money this week."
> "Sorry, mum can't afford sweets."
> "Sorry we have no fire today, the coal is low. Keep your coats on, chins up, don't let the scabs see you cry and remember, your dad is not fighting only for his job, but for theirs; and one day we will win."

The older children, who could understand why life had changed so suddenly and so dramatically, were able to appreciate some of the positive things to come out of the strike. They quickly began to learn the difference between 'values' and 'valuables'. Rita Abbott from Calverton explains:

> Money used to be all-important to us and to the children. Now we have other things, more important, there's a lot to compensate. We have new friends, from all over the world, people helping us, we share what we've got and our kids know why we're doing it all. We've got pride in what we're doing. You can't buy that at the shop, can you?

But despite this courage, and a determination to make the best of things, pressures increased as the weeks and months went by. There was constant anxiety about people being arrested or injured on picket lines, about those held by the police or given punitive bail conditions, about forthcoming court cases and possible imprisonment. For those miners convicted, there was also the prospect of being sacked, of not having a job to return to at the end of it all. When men returned from a picket the first questions in the strike kitchen would be: "Anybody missing? Anybody lifted? Anybody hurt?" An injury to anyone in the strike was felt by everyone. Women feared, too, for themselves, every time they went on a demonstration or a picket. They dreaded their own arrest, injury or imprisonment, but refused to be intimidated by their fears.

Treatment by Social Services was particularly distressing, though some workers in local offices were as helpful as they could be, and were often embarrassed by the things they had to do. Coal sheds were examined to see if there was a genuine need for coal for children; pantries were inspected when people claimed special payments; worn children's shoes and clothes had to be produced before replacement grants could be given. Again it was worries about the deprivation of their children that forced women to undergo this humiliation.

> We've always been able to provide for our kids, never asked for anything before, now they treat us like beggars. I really hate it, but it's our right and we have to have shoes for them or they can't go to school. I'm not letting the kids suffer, it's not their fault we're on strike.

Over the period of the dispute, inevitably, shoes, sandals and wellingtons became too small; first summer dresses, then winter coats became too short; trousers came half way up growing legs.

> We don't care for ourselves, we can manage with what we've got, but the children are growing fast and they have to be clothed. This is what makes you wonder if you're doing the right thing. We do know we're right, the strike is right, but the children have a right to shoes.

As savings dried up, debts began to pile up. Mortgage repayments, rates, rents, hire purchase repayments could not be met. Items of furniture were sold or repossessed. Some families had help from friends or relatives. And cases of severe hardship were usually helped from the fund administered by Henry Richardson and Pam Elliot, or by the pit support group itself. 'Holding' payments for gas and electricity kept the supplies on, but there was always the worry of not being able to meet even these essentials.

Then there was the agonising prospect of comrades and relatives returning to work. The women activists in particular felt acutely the shame and misery of sons, brothers, fathers, uncles, even sometimes husbands, breaking the ranks.

Though they insist that they never went short of food, tensions could not fail to have been increased by the lack of varicty—extra stodge to take the place of fresh fruit and vegetables. Every vanload of food seemed to contain tins and tins and tins of baked beans, unless it came from abroad, and then there were all kinds of new foods to be sampled: "I'll never eat bloody beans on toast after the strike, they're coming out of my bloody ears."

But for all their personal worries, hundreds of women threw themselves into the work of the support groups and maintained their commitment to the bitter end. Looking at the women involved is like looking into a kaleidoscope, seeing all their different qualities, experiences, attitudes, priorities, financial circumstances and personal relationships. The richness of this individuality combined to give a wide range of contributions and every group worked in its own way, drawing on the unique qualities of the women who formed it. Some women, such as Ida Hackett, had a long history of political involvement; others, such as Gwen Macleod and Toni Bennet, were experienced trade union activists. A few had been involved in the miners' strike of the 1970's. Rita Abbott asks why everyone was so surprised at the women's involvement this time:

> We did it in 1972 and 1974, not so much of course, but the work of 1984/85 was a follow-on of what we did then. Not so many women were involved, true, but we had a taste of what we could do if we had to.

Yvonne Woodhead, Marie Price and the women of Harworth were steeped in the miners' struggles, through fathers, uncles and grandfathers. They knew the story of the Spencer union and the shame that so many miners felt that it had brought to the county of Nottinghamshire. Most of the women were vaguely aware of the policy of attacks on the coal industry and of the build-up of nuclear energy as an alternative to coal. But for the majority, knowledge of trade union or political organising was second hand, gained through the experience of husbands, brothers and fathers.

Only a few of the women who came together to form the support groups were friends before the strike. Some were passing acquaintances but the majority were complete strangers. Through working together so closely, friendships grew and became firm. For most of them it was the first time they had experienced collective activity with other women or been involved in group decisions. They had to accept that their own ideas might be rejected, sometimes after long consideration, sometimes out of hand. They had to argue their own or someone else's case, or propose an alternative, and they had to defend their actions. They learned about the conduct of meetings, about such things as agenda and minutes. Some took to it all with an amazing ease, others found it more difficult. Some thrived in the cut and thrust of arguing and debate about the progress of activities, others found the conflict difficult to handle. Looking back at their early efforts, they are all amazed at how quickly problems were ironed out and things fell into a fairly smooth routine.

Many did not realise what strike action would mean until they had committed themselves to it. They feel that they were caught up in a kind of whirlpool, carried along on a fast-flowing current over which they had very little control. Who could go to speak in Bristol, Liverpool or Birmingham—to teachers, firemen or railway porters? Who could go on holiday with a group of children for a week—to Devon or Peterborough, to Holland, Germany or France? Who could help run a stall at the GLC? Who would speak at a fringe meeting at a women's conference? Who would be delegate to the central group? Who would get up at four o'clock on Friday morning to do a big women's picket twenty miles away? Whose car was roadworthy enough to pick up next week's provisions from the supermarket? No-one ever knew what was coming up next. Whatever it was might be as soon as the next day and there was often little time to prepare. But day after day, month in month out, there was always someone willing to do the work, even when, towards the end, numbers fell and women were desperately tired. Pauline Radford and Doreen Humber were involved with an Australian journalist, Lynn Beaton, who was writing a book about them:

> She would wait until we came home from the strike kitchen, or from the shops, a meeting or a rally, and she would want to do an interview. We just wanted to get our heads down for an hour before the next job. We wished she'd go away and give us some peace. When I think back over the strike, I was dead beat from morning till night, we all were.

The period of the strike was a frenzy of activity from beginning to end. There was little time for debate or reflection on how things were being done, particularly in the early months. Women were put into positions where they had to learn quickly, where their individual and collective strengths had to emerge very rapidly, where resourcefulness was the key to success. And they had to learn to trust each other and rely on each other. This trust and reliance bound them together and supported them in the bad times.

Due to the special features of the strike in Nottinghamshire, the women were given almost complete autonomy to organise and work as they chose. With the exception of Henry Richardson and Ray Chadburn, the Nottinghamshire Area officials were opposed to the strike, and apart from administering the official fund from Berry Hill, played no part in organising support. At local level, some pro-strike branch officials organised the Nottinghamshire Miners' Forum, but this was independent of Nottinghamshire Area officials and the women were left a free hand to get on with things as they thought fit. This freedom, and the way in which it allowed the women to develop both their own skills and a very effective support network, meant that they rapidly gained the respect and acceptance of their male colleagues. From the start, the women's coordinating group, the Nottinghamshire Women's Support Group, was represented at meetings of the Nottinghamshire Miners' Forum. (Usually by Ida Hackett, Pam Oldfield, Jackie Naylor, Liz Hollis or Tina Yemm.) At first the men were inclined to dominate the meetings but they soon found that in the women delegates they were dealing with strong and determined people who knew what they wanted and were not going to be controlled or directed by men. Pam Oldfield says:

> We knew how to organise and plan our work and we had already worked it out before we went to their meetings. We wanted to hear their plans and tell them about ours so that we all knew what was going on. At first they wanted to hold us back. We listened to them but when we were sure we were right we just went ahead, did it in our own way. They had to learn to listen to us, it was hard for them but they had to do it.

Ida Hackett thinks that at first the men felt threatened by the women's capabilities:

> It was a new experience for men to see women working in this way, handling enormous amounts of money, speaking about the strike to huge audiences all over Britain and overseas. Of course we listened to their suggestions and comments, but we took responsibility for ourselves, that was what they found so hard to accept at first. In time they saw how determined we were, and how capable, and they had to listen to us, even sometimes to look to us for help. We had arguments and disagreements, but we got through them and we were on the same side, weren't we?

So the Nottinghamshire women had the freedom to do what they thought was needed and to respond to emergencies without having to go through Area or local officials. Of course, coupled with this freedom was a large amount of responsibility. Women had to fill the roles of speakers, fund-raisers, legal and welfare rights advisors and press officers, on top of running the strike kitchens and organising food parcels. In other areas far more of this work was done, or at least organised, by union officials.

Welfare rights was an area in which women were particularly active. Most families were claiming benefits from the DHSS, Social Services and Education Authorities for the first time and were meeting barriers at every level. Claims were repeatedly returned for more information, were lost or redirected; payments were delayed, often for weeks. Many women felt that they were being deliberately humiliated and that this treatment was a calculated attempt to drive the miners back to work. The establishment at Ollerton Strike Centre of a Welfare Rights office gave enormous assistance and relief as experts came

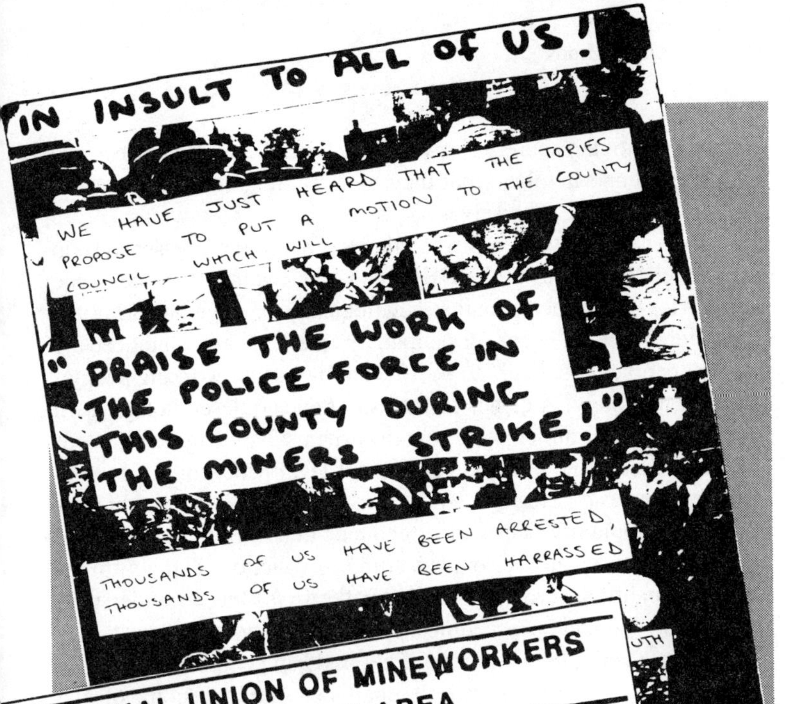

NATIONAL UNION OF MINEWORKERS NOTTINGHAM AREA

OLLERTON LEGAL ADVICE CENTRE

NEWSLETTER

26th July 1984

The Advice Service on Welfare Benefits is getting under way. We now have contact people in 14 of the 27 Womens' Action Groups in the Area and four regular sessions (in Clipstone, Hucknall & Linby, Welbeck and Mansfield) as well as the advice being given in Ollerton/Bevercotes Welfare.

If your Group doesn't have a contact person yet or wants to start a session then contact the Centre on Mansfield 862800.

If you are a "contact" then you will also find with this letter an updated version of the "Idiot's Guide". Read the old one, on the greensheet, and compare it with the new one on blue. There are one or two corrections and quite a few additional bits of information. Then throw the green one away so no-one gets confused.

We have had several queries about endowment mortgages. we will have an answer on them in a few days. Meanwhile people should remember that if there is real threat to someone's home you should contact the Cente immediately. we are already dealing with two threats (one from the NCB and one from a Finance Company) and look like winning tem both.

Contact people should now have:-
1. "Idiot's Guide", Edition 2, on blue paper
2. "TUCRIC" Booklet, with the orange cover
3. List of addresses for Notts.
4. Sheet about Family Income Supplement
5. Sheet about Notts. Educational Benefits
6. Sheet about Notts Housing Benefit.

If you are missing any of these contact Ollerton.

Finally, for those groups that have not got moving yet; it can really be worth while. We estimate that we have found some money for over half of the individuals and families we have seen. Sometimes it is only a few pounds but there was one who could get over £45 a week extra and several over £30.

from all parts of the country to give their advice. Some women became so interested and involved that they took training courses and were able to give advice themselves.

When necessary women formed themselves into a pressure group and lobbied the offices of the DHSS and other welfare services. One salutory lesson learned was that new legislation, often attacking the living conditions of working people, frequently goes unnoticed by the very people most likely to be affected by it; that working people, either by not knowing or not caring, so often acquiesce in their own oppression, never anticipating that they themselves will be affected.

Women also came to understand about legal proceedings, largely from being arrested themselves, or from the arrests of miners in their family or community. Through attendance in court, helping prepare statements with legal advisors, listening to solicitors explain things like bail conditions, women began, for the first time, to understand legal jargon and legal proceedings. As they realised how unjust most of the arrests were, how unfair the fines and bail conditions, and how few solicitors were prepared to challenge the decisions of the magistrates, women quickly became derisive of so-called 'justice'. I was often in court in Mansfield to hear proceedings against miners and others arrested during the strike and realised the disdain they felt for the 'legal system'. We tried to remain sitting when asked to stand and were often threatened with being turned out of court. We hummed "Arthur Scargill, Arthur Scargill, we'll support you ever more" in court one day and were ordered to be quiet. Not one of us had anticipated that we could ever behave in such a way but we gave each other confidence.

In order to be fully capable of speaking at meetings and to the press about the issues of the strike, women had to understand the origins of the dispute and keep up with local and national events. Never before had the politics of coal been so much a part of their lives. Men had rarely talked to women about their industry before, preferring to keep this as men's talk in a man's world. The women needed to be able to explain the special situation in Nottinghamshire and to give a true picture to set against biased media coverage. They recognised the war of nerves being waged by the Government and the National Coal Board in an effort to drive or seduce men back to work—the initiation of talks which would inevitably break down; the publicised offer of bonuses, extra holiday pay etc. which turned out to be nothing more than what men were entitled to anyway; the characterisation of miners as violent thugs. Everyone involved in the strike was enraged by the lies and misinformation being fed to the public and by the suppression of the facts. They hated the deliberate vilification of the national union leaders, particularly Arthur Scargill. This anger and frustration could so easily have sapped their strength and left them debilitated and depressed. But there was no time for that, and the work went on.

Hundreds of letters were written to supporters outside Nottinghamshire. Some women found this difficult at first, but it was a worthwhile experience, finding the words to express how they felt and what they were suffering. Their recipients welcomed this personal link with the coal communities. The letters were read out at meetings of activists all over the country and abroad, conveying thanks for money, food, clothing, wool and things to be raffled, and describing the latest developments in the area. Women who took on the job of letter writing often sat up into the early hours dealing with the masses of correspondence. Those who had written very little since they left school now found that their stories flowed easily, painting the picture of need and gratitude.

Most of the support groups were fairly self-contained, and problems between them were rare. One that was never really resolved, though, was the matter of twinning, or group adoption. A group in a pit community might be linked to a trade union branch, a political party or some other organisation, or even with a whole town. The idea was to concentrate exclusively on one pit, so as to ensure a regular flow of money, food and other support to that particular group. Early in the strike, two Nottinghamshire support groups, Blidworth and Ollerton, realised the urgency of organising to get money into the county, since it seemed highly unlikely that any Nottinghamshire Union funds would be made available. As it turned out, each group received £30 throughout the strike from the official Area Hardship Fund, but this was a tiny fraction of what was needed. The men and women of Blidworth and Ollerton made their earliest links with Cambridge and Norwich, contacting all organisations there and asking for help. The response was immediate and generous beyond all expectations.

But as Ida Hackett explains, there was another reason why support came into Blidworth and Ollerton so early and on such a scale.

> It was a way of expressing special sympathy to the two groups after the police occupation of Blidworth had been shown on television and also after the death of David Jones, a Yorkshire miner, on the picket line at Ollerton. There was so much media attention to these two events that people throughout the country heard the names of the two villages and in the early months they became better known than other pit villages. The publicity brought a great deal of money to Blidworth and Ollerton.

Secured as it was, early in the strike, this support meant that the two groups were almost self-sufficient for the duration of the dispute. The early contacts were strengthened by reciprocal visits and individual families came to know each other, learning a great deal about each other's work, lifestyles, attitudes and recreation. Friendships blossomed with mutual understanding and appreciation. Men and women went to their twinned city, speaking and collecting or for a short break from the pressures of the dispute. Supporters came to the pits, joined the picket lines, got arrested, took photographs, wrote articles and reports. They attended rallies and meetings together in other towns and cities, often marching together under each other's banners. This personal contact and solidarity formed the bedrock of support in groups like Blidworth and Ollerton.

When other groups in Nottinghamshire, later organised and a little slower in making contacts, saw the comparative well-being of the strikers and families in Blidworth and Ollerton, some felt resentful that they themselves were not so fortunate. Though they admitted that the support there was of those communities' own making, the result of their initiative and hard work, they sometimes felt that it was unfair. But the problem was not an easy one to resolve.

Even had the people in the two Nottinghamshire support groups been willing to sever or weaken their links with their twinned cities and ask for resources to be diverted to the central group or to other pit groups, it was highly unlikely that the people of Norwich or Cambridge would have readily agreed. They valued their personal contacts too highly and took great pleasure from knowing exactly who was benefitting from the money they raised and the food they provided.

It was this identification with 'their' pit and 'their' people which allowed people outside the pit areas to feel that they were really a part of the dispute.

What did happen was that Blidworth and Ollerton required no support from the central funds, unlike most other groups, thus leaving more for those in need. They directed would-be supporters to groups that were badly off, and when they could, they donated money to the central group by speaking on its behalf. When my own group in Newark, for example, approached Blidworth because we had three miners on strike from that pit, we were asked to give all our support to Bilsthorpe and Rainworth, two groups in great need at that time.

And so it was that twinning became the practice for almost every group. In the light of events it is difficult to see how one vast group could have operated in Nottinghamshire, responding to local or individual needs in the way in which the local groups were able to do. It is unlikely too that the vast amount of help which came into the county would have done so otherwise. The other loss would have been the emotional support which was so much a part of the 'adoption' process and which sustained the men, women and children right through the strike. The association of miners and their families with people hitherto outside their experience and the mutual understanding which developed between the two groups would have been on a far more limited scale.

Despite many attempts in Nottinghamshire to iron out problems of unequal aid, especially on the part of the central group officers, inevitably some inequalities remained. However, when strikers and women supporters went to areas like Yorkshire which had so many strikers in comparison, they realised that not only were there inequalities between groups in this county but also between areas. From time to time, and especially near the end when things were so bad, people from some groups in Nottinghamshire tried to help their comrades in other areas to ease the terrible privation which the winter brought. This further cemented the feelings of being together in one common struggle.

At a local level income fluctuated from week to week and many of the groups were never sure how long the flow would continue, even from the staunchest supporters. As the strikers' own resources and savings dried up, there were increasing demands on the money available. More and more people relied on the meals provided by the kitchens and the food parcels available for each family. There was particular concern for the single miners who were often totally dependent on the resources of the support group. Yet everyone who came to the kitchens expected the women to be smiling, cheerful and optimistic, to put on a brave face no matter what was happening personally, locally or nationally.

Usually tension was eased by the laughter that little incidents, such as the packing of food parcels, caused. In the Annesley and Newstead group Chris Simpson somehow never got the items in the right sequence: "Beans, soup, cornflakes, jam, whatever it was she always missed something out." The others treated it with good humour, had a good laugh together and got on with the job. Sometimes the fun of just being together was enough, even of being jostled in the kitchen. Maggie Ryer and Jane Marsden from Sherwood remember the arrival of bags of second-hand clothing as a high spot. Every group recalls laughing a lot, even though they no longer remember what caused most of it. "It was the little things. And if we hadn't laughed, we should have cried. There wasn't much between crying and laughing sometimes."

Just occasionally the pressures got too much and a member of the central group, often Ida Hackett, was called in to help sort out problems which threatened to damage the work of the group. But that was rarely necessary, often problems which seemed impossible to deal with were resolved when women remembered that they all shared the same goals and that so many people were depending on them. I talked with Sue Petney from Blidworth one day about the widely-held view that women are catty. We came to the conclusion that this is attributed to women by men, but that women too often accept the characterisation without thinking what it really means; that women's differences of opinion are put down to their personalities, rather than their viewpoints, whereas men's differences are seen very differently. Unfortunately, in the relentless activity of the support groups there was rarely time to sit and discuss such things; even when women stopped for a cup of tea their thoughts were usually on more pressing matters. What is incredible, on reflection, is the harmony which existed in the groups, when one realises all that was against them and when one recognises the extent of the work they did to establish the massive alternative welfare support structure in Nottinghamshire.

In the communities as a whole, parties helped to keep spirits high. Though food and drink were always in short supply, often someone would come up with some home-brewed beer or wine, or supporters from outside Nottinghamshire would provide all that was needed to create a social atmosphere. Parties were a means of drawing closer together, physically and emotionally, of shutting out enemies and showing defiance. This was particularly important in an area where a striker could be isolated in a street full of working miners. Bright lights, laughter, singing, especially of strike songs, were meant to show high spirits and well-being; that there was no despondency and a great deal of support. The noise was mischievously intended to inconvenience and annoy 'scab' neighbours and often police arrived in answer to complaints. Elsie Henderson from Welbeck describes her silver wedding party on August 29th, 1984:

> The weekend prior to this, some of our striking miners and their wives and children had been the guests of Cortonwood miners. In general conversation it was mentioned that it was our silver anniversary the following week. Some of the Yorkshire people asked if they could come down and have a drink with us. We agreed that they should come and stay overnight. After tea we went to the Welfare Club. On the way we passed two policemen. I'm sure they recognised the Yorkshire accents. The men were wearing Cortonwood Flying Picket 'T' shirts, so they couldn't fail to notice them, and we knew we could expect trouble. After a drink at the Welfare, then at a pub, we noticed three police on guard outside. The men from Cortonwood kept inviting people back to our house, and though we were having a good time I knew we couldn't give them anything, not even a cup of coffee! Then people started to arrive with home brewed beer and wine, and food as well. There must have been about two hundred packed in our house that night.
>
> The party was in full swing at about two o'clock in the morning when sure enough there was a complaint from a working miner. There had been two minibuses of police drawn up outside our house and the Yorkshiremen had gone outside shouting "The miners united will never be defeated." The vans had driven away but as

> we went back into the house three panda cars arrived. We ushered everyone inside, closed the door, went down to the gate and stood leaning on it; myself, my husband and my son. We explained the reason for the party, and added that the noise on the street had been sparked off due to intimidation by the police. The sergeant asked if the noise, which we could barely hear, was the same as before and when we said "yes", he told us to carry on and enjoy ourselves. It was hilarious and we went inside. The party went on until about five in the morning. No-one will ever have a silver anniversary like we did and we will treasure the memory always. We often laugh as we imagine the headline, **Mass Arrest at Silver Wedding in Meden Vale.**

However, when we ask what it really was that kept people going in the face of arrests, bail conditions which denied all civil liberties, road blocks, mounting debts, increased shortages, misrepresentation by the media and the divisions in the community, the answer is the belief that the cause was right and the battle totally justified. There was also the certainty, right up until the period just before Christmas, that they would win.

Though a few of the women in the support groups had been involved in trade unions or political groups before the strike, by far the largest number of those active admit that before it began they were politically asleep. The women's liberation movement had nothing to say to them:

> We thought the women who went to Greenham Common were what the papers said they were, cranks and drop-outs, leaving their kids and living in the woods, dirty and scruffy.
>
> ★
>
> We knew that somehow women always seemed to get a raw deal but we had never really sat down and worked out just how this happened. We never worked out our needs and thought about abortion laws, divorce laws, low wages for women, lack of nursery facilities. We weren't bothered really about black women or about the treatment of prostitutes. It was all so vague to us, our own fault because we never bothered, we were all tied up in our own little lives, in our own little houses and villages.

But as women began to mix with a wide range of different people outside the coalfield, and as they experienced the full weight of the state against their own community, they quickly began to rearrange their ideas and examine their prejudices. In the space of one year, some women changed so dramatically as to be unrecognisable, even to themselves.

> Somehow, the strike changed our view and we looked out of a great big window and saw things for the first time. I can't believe how ignorant we were, and I'm really sorry now that we didn't care before.

Nottinghamshire women made friends with black and Asian women and began to understand how black people have suffered for years some of what the mining communities were suffering in the strike.

> We didn't know any (black people) before the strike, it's really knowing them that makes the difference. We watched the pictures of the riots and saw black people on the streets, fighting against the police. It's only now that we have police against us that we see things in a different light. We believed what the papers and the TV told us. Now we get furious because people do the same to us. But we were just the same once. Isn't it bad that you have to go through it yourself before you see things properly? If only we could all be together, all helping each other, then we really would be strong.

They made links between nuclear power and the future of the coal industry, between the use of the police on the streets of Northern Ireland and their deployment in their own villages.

As the dispute progressed and the strikers began to realise that they were being disciplined by a government determined to break them, no matter what the cost; as they began to look beyond their own treatment to the exploitation and oppression of others, they became avid for knowledge, ideas and argument, knowing at the end that they would make up their own minds in the light of all this new information.

Most women's experience in mining families, as elsewhere, is a sexist one. In a miners' world women were protected and provided for. They began to realise that this also controlled and restricted them.

In order to do the work of the strike, women had to push aside these restrictions and men were often deeply alarmed by what they saw, fearful of the long-term implications. They attempted to ease their worries by making a joke out of it: "We'll be all right when the strike's over, you won't have to go racing about all over the globe, you'll be home to cook my dinner."...."Me and our lass'll be right again in no time, just like we were before."

For a lot of men this was bravado, a kind of desperate hope that things would return to the same old domestic relationship. Maybe that will be so on a superficial level but attitudes which have changed will not revert to former acceptance of male dominance. Women now know how able and powerful they are in so many new spheres, and that knowledge will last for ever.

Strike centre kitchens became the focal point for political discussion, especially when supporters from a wide-ranging spectrum of backgrounds, both social and political, came to visit the various groups, as they did in their hundreds. Coming to express their solidarity and to see for themselves what was happening, they brought with them their ideas, their newspapers and their pamphlets. When there was time to listen, the women found it all very exciting and novel. Some made instant affiliation with one particular group, others preferred to listen to them all and show interest without making any particular commitment.

In some strike kitchens a television was available so that every item of strike news was viewed and analysed collectively. From this experience women learned what a powerful influence television is on our lives, especially when programmes are watched in very small groups at home where there is usually no-one to challenge the presentation or propose alternative views. In the same way, having cancelled regular paper orders for economy, they shared one copy at the centre. In the kitchen, groups watched together, 'read' the media together, and began to identify propaganda and indoctrination. They rejected the majority of the national newspapers, regretting that for so long they had subscribed to such material. They realised how gullible they had once been, how they had underestimated their capacity for understanding national and international matters and how their views had been formed by the media, manipulated by distortion, misrepresentation and omission. They learned about censorship and they became sceptical as they saw their own experiences in the strike presented with blatant inaccuracy and bias. I remember the women at Blidworth refusing to give an interview to one reporter from the 'gutter press' and showing him the door very forcefully. Other journalists seeking stories and wanting to do interviews were treated with extreme caution and for most the answer was firmly negative. Doreen Humber, with her usual candour, told me:

> At first we began to think of ourselves as star personalities, enjoying doing interviews for papers, radio and TV. We came up with a jolt one day when we had a woman copper in the centre asking for an interview. We never thought to check and we didn't find out till after. It stopped us in our tracks and we remembered we weren't stars at all.

But for many of those involved in the strike, the most dramatic politicising experience was the behaviour of the police. From early in the dispute police activity in Nottinghamshire was intense. Though other coalfields suffered this police involvement later, for Nottinghamshire pit villages it happened very quickly as mass pickets came from Yorkshire and elsewhere to try to bring out the Nottinghamshire men. Police were drafted in from every force in the country in an attempt to stop them. The miners were criminalised to win popular condemnation and weaken support for their cause. They had to be shown as brutes who attacked life and property indiscriminately, as ruthless and gratuitously violent. The police were shown, even when pictorial evidence discredited such portrayal, as impartial peace-keepers, protecting the 'democratic' rights of strike-breakers. In Nottinghamshire, closure of roads became a daily occurence, as did questioning of drivers and passengers, searching of vehicles, and taking of car numbers. People were prevented from moving about their towns and villages. Even children were turned back in their own streets, or questioned about their families' involvement in the strike. In Warsop Vale foot patrols of four policemen arrived every fifteen minutes:

> They stopped anybody they thought had anything to do with the strike, even asked the kids if their dads had gone picketing. They

came day and night, in vans, cars and sometimes on horseback.

Sharon Reed from Sutton-in-Ashfield recalls that they even got into the support kitchen:

"They were always prowling around, snooping on us, seeing who was there. They even put a guard on the gates sometimes. We hated them being anywhere near us." There were also stories of how nice the police were at first and how naive the women had been, disarmed by the good-humoured banter and the sweets given to the children. "They used to swop their badges, for strike badges; we're sure that this was to let them get on the picket lines later on and stir up trouble to get miners arrested." The police soon changed and it became obvious which side they were on. All pretence of fairness or impartiality was dispensed with and they clearly demonstrated the lengths they were prepared to go to. In places such as Blidworth they invaded private homes and gardens, trying to drive out the pickets who were guests of some of the families. In Cresswell children hung over the school wall watching the picket lines and police activity. Pam Richardson tells of children refusing to go into lessons after playtime when something dramatic was happening.

> The teachers just couldn't get them back into the classrooms. They used to shout 'scab' to the working men and sing strike songs. I think it was exciting for them. It's awful though now, they don't care a bit for the police. There's no respect like there used to be, not even for the village bobby, that's all gone now.

Now the police are seen simply as agents of the state: "How can we now go to the police for help? We have seen what they do to people, we've seen it for ourselves."

The contacts they made during the dispute led many women to question their own education and expectations. Before the strike few of them had had anything to do, on anything like an equal footing, with professional men and women. Gradually they began to relate with ease to journalists, writers, film-makers, artists, solicitors, teachers, lecturers and politicians. They began to look back over their own school experiences and realise how their expectations had been shaped and conditioned by the system. Lynne Dennet and Alison

Lovely from Church Warsop recalled how teachers had controlled their learning and limited their horizons:

> We didn't think we could ever be anything. Nobody ever lifted our sights. Kids like us didn't become doctors and solicitors or people in jobs like that, it was shops, factories and the pit for us. We never really learned what we could do; we've missed out all along the line in education. The system isn't for us."

Vickie Smails from Rhodesia speaks for the others in her group when she describes how her schooling failed her, suppressing rather than encouraging her abilities.

> I've got to make up for the education I missed in all the time I spent going to school. When I think what women like myself could have done if only we'd had the chance. Still, its not too late and there are things to do, and places like Northern College for us to make up for lost time. I'm going to make sure my daughter doesn't miss out like I did. But the system doesn't work in our favour. We have to try even harder to get anything out of it.

Perhaps for the first time, women began to realise that they had never had an opportunity to show their potential. The strike gave them this chance, a chance to prove their value to themselves and to others. In a perverse way the strike brought excitement, put them in the limelight, and gave new meaning to lives which had hitherto been routine, sometimes drab and often boring.

Tony Benn, MP for Chesterfield, has said of the women that they learned more political lessons in one year than other political activists such as himself have learnt in a whole lifetime. He has emphasised that others must now learn from them. Without a formal structural base, without such things as standing orders, rules and constitutions, the women of the coalfields launched a massive political education campaign, publicising throughout the country and abroad what was happening in the coalfields and its implications for other workers. They swept aside restrictions, challenging and even overriding authority when it stood in the way of their immediate needs, going right to the heart of the matter. Their insights were gained through direct experience. They learnt all about the subtle as well as the crudely obvious controls that exist in our so-called democratic country. They learned and were able to teach others about the coal industry.

Everyone who worked for the strike was changed to some extent by the new things they saw, the new people they met and the new places they visited. But most of the women would not want to change their own lifestyles. It is their attitudes which have altered, the way they interpret issues and events, and to some extent the way they view themselves. They remain fiercely loyal to their own way of life and in particular to their relationships with their husbands and sons. Having played their part in the strike with their own kind of dedication and their own integrity, they resent any suggestion that they should become totally different women.

> We met feminists and they frightened us, nearly as much as they frightened our husbands. We could never go that far but we have taken bits of their ideas. We don't mind being called 'ladies' instead of 'women', we don't see that it matters. All that the feminists did for us really was to push us a bit further in being confident. But we had to be confident, didn't we, or we'd never have got anything done.

Sandra Taylor from Welbeck and Sharon Reed both agree that the biggest difference is having gained the confidence to learn: "It's no good being confident if you don't know anything, you just get made to look a fool. You have to keep listening and asking questions, then you know as much as anybody in the end."

Many of the women have become active in the Labour Party, not only learning but also questioning the way in which the party is run. Sandra is now a Parish Councillor in her village and an activist in her constituency party of Bassetlaw. Pam Oldfield and Jackie Naylor in Mansfield, Gwen Macleod and Sharon Reed in Ashfield, Doreen Humber, Pauline Radford, Betty Savage and Yvonne Woodhead in Sherwood, Vickie Smails and her friends in the Rhodesia group—all are in the Labour Party, attempting to slash their way through the bureaucracy which so often seems to inhibit action. They become impatient with the men who seem to be hide-bound in constitutional matters. Dissatisfied with the attitudes of some local Labour MP's and Councillors both during and after the strike, women are now seeking election as Labour Party officers in branch and constituency parties, as well as nomination as candidates in local elections. They want to inject into the Labour Party some of the urgency and action which they brought to the problems of their community on strike.

Many people have applauded the efforts and achievements of women throughout the British coalfield in the dispute, but perhaps they should not have been so surprised. Women everywhere had proved time and again that they have great energy, spirit and ability. Throughout the world, women are joining in sisterhood to fight against injustice, oppression and force. The women of the coal strike were part of that fight.

NOTTINGHAMSHIRE WOMEN'S SUPPORT GROUP

Affectionately known as the 'Central Group', or the 'Umbrella Group', the Nottinghamshire Women's Support Group was the delegate body of the women's support groups throughout the county. It grew out of a meeting early in April 1984, when four people in Mansfield got together to discuss the strike and what needed to be done. Julie Wilkinson, Ida Hackett, Ernie Daglish and Andy Miller all anticipated that it would be a long struggle and felt that the first thing was to get the involvement of men and women from all the mining areas. News of this first meeting, which was held in Mansfield Community Arts Centre, was spread by word of mouth. The early initiative was well-received and further meetings were called.

Members of the small committee elected at the first meeting began to visit the mining communities, and in particular the women, giving advice on getting support, setting up kitchens, organising food parcels and contacting the DHSS and other social services departments. Thus was set in motion the very successful network of welfare care which sustained those involved in the dispute. As the group developed it became increasingly a women's organisation. Both Ernie Daglish and Andy Miller withdrew and their positions on the committee were taken by women. There was no formal constitution and in July the Committee was extended to include more women. Weekly delegate meetings were held alternately in Nottingham and Mansfield to cater for the groups in the north and the south of the coalfield. In Mansfield, meetings were held at the Community Arts Centre, the Mason's Arms, Charters Arms or Mansfield Library; in Nottingham they were usually at International Community Centre. At these meetings Coop tokens, clothing and anything else available was distributed. In order for the central organisation to function effectively it was necessary to have a number of contacts in each local support group, to whom weekly newsletters and agenda were mailed out. These contained notices of rallies and meetings, and any other information of relevance to all the women. In this way it was intended to have full dissemination of information. Details of income and expenditure were included so that everyone had a clear picture of the financial position of the central fund.

Each local group had its own autonomy but the central group existed to help out with things which they could not handle. Major problems like acquisition or maintenance of premises were dealt with, as were problems of shortage of money. But there were other problems of individual differences which needed a 'neutral' view or a different perspective. Ida Hackett was often the one who went directly to the women to help, and her opinions and advice were well respected. Some local groups were particularly vulnerable and dependent, whilst others were self-sufficient and often isolationist. Different levels of capability emerged in different groups. Some progressed from a position of dependency on the central group, through self-sufficiency, to the ability to help fund the central group.

The central group was heavily involved in the organisation of speakers to go out fund-raising throughout the country and abroad. It dealt with collection and delivery of food from the National Union of Communication Workers, National Union of Railwaymen, the General and Muncipal Workers' Union, and of food tokens. Originally the central collection point was the Friends' Meeting House in Mansfield, but later it was moved to a house in High Pavement, Sutton-in-Ashfield. Women from the group worked through the NUPE office in Nottingham to organise holidays, coaches for rallies and meetings, women's pickets, collections on streets and in pubs, in towns and cities all over Britain, and food and presents at Christmas. The group also made fund-raising materials available to individuals and groups supporting the miners. It arranged stalls at conferences and festivals and it tried to obtain permits for women to collect locally.

Whereas groups in the solid mining areas had few difficulties in obtaining stalls on markets and in the streets, the Nottinghamshire women met with obstruction and difficulty wherever they applied. One instance in which the central group was involved was the attempt to obtain a charity stall in Mansfield for use by the Clipstone Support Group.

The idea for the stall followed the successful May Day Rally organised by Mansfield District Council, at which Arthur Scargill was

Nottinghamshire Women's Support Group Committee, 1984; Back: Audrey Mulholland, Rita Abbott, Liz Hollis, Margaret Brown, Front: Pam Oldfield, Ida Hacket, Jackie Naylor

Paul Dorran

HERE WE GO!

BULLETIN OF THE NOTTINGHAMSHIRE WOMEN'S SUPPORT GROUPS. No 4 7 th JULY

THANKS!

The Nottinghamshire women's action groups welcome the solidarity and support of the people of Sheffield, East Anglia, London and elsewhere who are joining in the great 'FOOD FOR VICTORY' celebrations on July 8th. South Yorkshire is already famous for its cheap buses; but 20 double-deckers will be free to all passengers on Sunday when they'll set off from Sheffield in a convoy, led by the Lord Mayor, and crammed with food for the Notts. soup kitchens.

At 10 meeting points through out the county the women's support groups will be welcoming the buses; some are organising fun football matches and other events; all will have a chance to chat with so me of the many people who believe the miners' cause is right.

Donations of food and money, w ch are being collected in Sheffield throughout the preceeding week will be unloaded in the villages; striking Notts miners and their families will then ride on to Forest Fields in Nottingham, where a carnival is planned on the Goose Fair Site, with free food, drink and entertainment. There are international links too... Turkish workers from London are donating thousands of kebabs!

The idea for the event came from our sisters in Sheffield women's action group and was taken up with enthusiasm by Sheffield Trades Council. It's good to know that Yorkshire people recognise the courage of the Notts. strikers. Our N.U.M. gala used to be a great day out; but it had to be cancelled because of the disunity in the area. The women's groups are helping carry on the good traditions of Notts.; the tradition of Harworth miners who broke the Spencer union in 1937 and brought the area back into the National union; the tradition of the miners' wives who pawned their wedding rings, pinned the tickets to their blouses and marched on Mansfield work house during the '26 lock-out, refusing to be starved back to work.

'FOOD FOR VICTORY' will give us a breather in the fight to save our communities, despite those men who aren't prepared to fight for their jobs. Thanks to Sheffield!

Lesson to the scabs

Sad farewells were said as the Sherwood ladies left Kent after the weekend spent down there for the Kent Area women's demonstration on Saturday June 23rd. 2,000 women from areas all over the country marched through Dover and up to the castle. The speakers included women from Kent, Yorkshire, Wales, Loughborough, Nottingham, Young Socialists from the Labour Party, Black workers and a representative from the French Union of Mineworkers. The message from all was the same - 'United we will win'.

The Sherwood group who represented Notts. were treated by the Kent families like members of their own families. The hospitality shown to us by these mining communities was unforgettable. We were taken to visit Snowdown pit.

Cont.

Description of "Food for Victory" event in *Here We Go*

the main speaker. Miners and trade unionists from many parts of Britain had come with their banners to make this a special day for the strikers, and thousands of people took part. The support groups were given stalls free of charge to sell their goods and raise funds. The central group produced a newsletter, *Here We Go*, and many links were made, with invitations to send speakers on fund-raising visits.

It was the success of this event which prompted the approach to Mansfield District Council for the use of the stall. Permission was given and Eileen Ripley, Ann Chapman, Isabel Duncan and Ida Hackett collected. It was a great success. However, this success was short-lived as the council subsequently passed a proposal unanimously prohibiting use of the stall to all but registered charities. Allegations were made that trouble at the stall had led to its being closed down by the police. This was totally untrue and the women refuted it vehemently. However, protests were to no avail and the ban remained.

Since at this stage support from within Nottinghamshire was very limited and nowhere near meeting increasing needs, it was necessary to look outwards and try to tap all sources. Ida Hackett wrote an appeal to *The Morning Star* and within a month readers sent £1,182 into Nottinghamshire. Many readers initiated support groups in their areas, some of which continued to send regularly for the duration of the strike.

It was through links made in this way with Hackney that Ida Hackett first met Danny Phillips who eventually came to Ollerton and worked so hard to get the best possible welfare rights for the striking miners, women and children.

Relationships with the men's central strike committee, The Nottinghamshire Miners' Forum, were usually good, though there were periods of tension. At first one delegate from the central group, and later two, usually Ida Hackett and Pam Oldfield, attended each meeting. It was obvious that this was initially seen by the men merely as a token gesture, which did not satisfy the women delegates. There was disagreement over collection 'territory' and funds, with accusations of 'trespassing'. Finally the women gave an ultimatum, they were to be accepted as equals or there would be no participation at all. The men were forced to agree and came to accept the real value of the women and their work. Their attitude to the women changed as they experienced their ability and determination. Finances were kept separate, but towards the half-way mark there was some mutual aid.

No woman who was involved in work for the strike was barred in any way from involvement in the central group because she was not directly related to a miner. In other areas, outside Nottinghamshire, there was great admiration for the way in which the group functioned, for the way in which the women overcame the special problems of the area and for the high level of morale in the groups.

The meetings were well-attended and very much enjoyed by the women delegates. It was a chance to hear what was going on elsewhere, to get ideas, to see old friends and new, to cement relationships made on marches and at meetings out of the area, to see how everyone was getting on and to feel bonds of solidarity.

Pam Kenny from Gedling, working in a group which was very small, found the visits to the central group meetings particularly helpful.

> I really loved them, I only missed them when I hadn't any transport. It helped to keep me going when I saw all the others, they were always enthusiastic and had lots of things to talk about. We knew we could get help if we needed it, if money ran short, that was good. I suffer from agoraphobia and it is hard to get out, but I managed it all through the strike, I had to.

There were often visitors to the meetings, always vouched for by a member of the group—a photographer, a film group, students wanting to make a video, women from support groups out of the area. Occasionally Henry Richardson came to make sure that the events of the strike were fully known and understood. This was particularly important when complicated legal actions were being taken against the NUM. Women had to make sure that they really grasped it all. When they went to speak at meetings they had to be able to explain the strike and respond to questions from people outside the dispute. The knowledge they acquired made them very confident.

I attended many of the meetings as a delegate from my group in Newark. I always felt proud to be in the company of women who were working so hard and learning so fast. I saw women develop from saying nothing at first to standing up and speaking their minds with force and clarity. They seemed to me to say things so directly, with an economy of words which I envied.

One young woman who was always at these meetings was Liz Hollis, bringing reports from the NUPE office. She usually arrived rather late, arms full of papers, lists, notices, leaflets and messages. She always appeared to be in a flap but that was because she did so much. Liz, or 'dizzy Liz' as she was affectionately known, ran the NUPE office, worked on the collecting stall in the city centre, went picketing, marched and rallied, spoke at meetings, collected in the pubs and clubs and drove her battered van all over the country. She was full of fun, bringing her good-natured brand of humour to everything she did. Everybody loved her, though she sometimes forgot things, or was late or elusive; everyone recognised what a massive contribution she was making.

I remember the meeting when she was trying to get some women to go with her to Liverpool: "We're going round the pubs, collecting, come on, who's coming with me? I'll look after you, you'll be all right with me." The prospect was daunting, but somehow her confidence and determination was infectious and, sure enough, women volunteered. Anybody who wanted a bit of fun would be sure to get it with Liz:

> My old van saw some mileage before it finally decided it had had enough and was packing up. It was always breaking down, leaving us stranded. I drove it to take kids to ports and airports for holidays abroad. I drove it full of women going fund-raising, and I took men all over the place. I used it for logging, for coal, when we had any, to fetch food and toys; you name it, my old van carried it.
>
> I was unemployed when the strike started. It seemed a good idea to get busy for the miners. The more I did the more there was to do. I never had such a year, it was wonderful. Even when things went wrong it always came out right in the end.

Ida Hackett recalls one example of Liz having fun. It was just before Christmas and there was a school full of toys to be organised and distributed. Ida arrived to find Liz with just enough room to stand, absolutely surrounded by toys of every size and description, almost exhausted and getting fraught by the size of the task:

> It was obvious she'd just about had enough, she was worn out. Then she found a pair of roller skates. Before we knew what was happening, she had them on and was out of the door into the school yard, whizzing round on these skates. It was pouring with rain but round and round the yard she went, laughing and shouting like a child, really enjoying herself, wet through. She was letting off steam, getting rid of her frustration. That was Liz to a 'T', up to all sorts of things, laughing even when things were going wrong. I think she helped us all to be cheerful.

She could have taken a job during the strike, she was offered work but felt that she was too involved, too many people depended on her and she couldn't let them down. She turned down the offer and stayed with the strike. A vicar's daughter, with no direct relationship with the mining community or the mining industry, she gave the whole of her energy and concern to the dispute and the needs of the striking men, women and children of the Nottinghamshire mining communities.

Two other women central to the group were Ida Hackett and Pam Oldfield. Ida Hackett was 70 during the strike, yet she had the energy of a woman half her age, or so it seemed to the younger women. There was to be no end to the work she did. She was probably one of the very few who knew from the beginning that this was likely to be a long and bitter struggle. She anticipated that the government would be determined to bring about a different conclusion to this strike after their humiliating experience at the hands of the miners in 1972 and 1974. She had also read the Ridley Report and had some idea of the government's plans for the coal industry:

> But I never anticipated what I should do in the strike, there were so many things. I can't tell you all of them. I went picketing regularly in all weathers. I marched and went to rallies and conferences and I spoke at meetings all over the county. The work really started from the rally on May 14th in Mansfield. The Nottinghamshire one page bulletin was distributed to all the women's groups

Peterborough
PE4 ~~6SJ~~ 6 8J.
May 17th 1984

Dear Ida,
Enclosed cheque for £20 for the
Miners support fund which you wrote about in
the Morning Star. I will send this monthly
for as long as it is needed, I can only afford it
by giving up smoking, so it will do me a
good turn too. I have a box of womens and
childrens clothing which I will try to get
delivered to you and will look out for jeans
at our local jumble sales. We have had a
mtg (c.P) with a Kent miner speaking and
collected £60 for them. The local L.P. is doing
nothing so far, only one or two collecting food.
which is going to Kent, but we are working
on it. I wish your committee every
success in your efforts and may the strike
soon be over with victory for our people.

Yours fraternally.

there and Newham immediately asked me to go down to speak. I went alone in June, I can't tell you how many meetings I spoke at, large and small, and all the money was for the central group, to help the small groups being set up in the pit villages. That rally was the start, but it was a pretty hair-raising day for me in some ways. I'd arranged to meet about 50 or 60 Yorkshire lads in a corner of a car park to fix up where they were staying overnight. We got all that organised, we had the rally and the meeting outside when Arthur Scargill spoke, and I was just settling down to a cup of tea with Ernest, my husband, when the phone started to ring. It kept ringing again and again, more and more Yorkshire men were asking where they could stay the night. Our flat was choc-a-block, the car park outside and the street full of cars. I was going round at 10 o'clock that night knocking the doors of people I knew were sympathetic, asking if they had spare beds, or places to put sleeping bags. Ernest didn't know if he was coming or going, making cup after cup of tea. I made so many contacts from that rally. One of the very best was with the Union of Communication Workers. They were wonderful. Do you know, they sent £1,500 of food every week from the end of May until one week after the strike. They sent money, toys, and poultry at Christmas and they offered short stay rests for people from the strike with all expenses paid. I don't know what we should have done without them and all the other people who helped so much.

I couldn't have done what I did without Ernest. He was secretary, housemaid and cook. He did what housework got done, I was a lady of leisure in that area. He took all the telephone calls when I was out. I'd come back, often tired out, and there'd be a list, all written down with times of phoning, who phoned, the message, whether to ring back, everything recorded.

The only meeting I refused was in America. Ernest was ill and I wouldn't go and leave him, though we had a row about it as he wanted me to go. I turned it down and I'm glad I did because he died before the women came back. I met so many people in the year, I went to so many places. I've always been active with campaigns all my life, so it wasn't so hard for me as for the other women. For them it was a complete change in their lives.

Pam Oldfield was a founder member of Sherwood Support Group, then became more involved in the central group, taking over as secretary from Julie Wilkinson.

I was with Sherwood for about four months and I went out speaking for this group. My first time was in London and I drove there, the first time I had ever driven in London. I had an address on a piece of paper, it was the AUEW. I just drove round and round until I finally found the place. I was nervous to start off with and I got more and more nervous and flustered thinking I would never find it. You can imagine the state I was in when I finally arrived. It was hilarious, I knocked a glass of water over the chairman's knee; he was soaked, but he just smiled. Bless him, he must have known how nervous I was.

I liked the work as central group secretary, I'm better at administration than at public speaking! But even so, I had to go speaking if there was no-one else to go. On one occasion I went to Kent. The invitation was from the women and I thought it would be a meeting organised by them. When I got there I found it was a Kent miners' rally and I was a platform speaker! The bones went out of my legs. I thought I should die. I wanted to die. To cope with the nervousness I imagined all the men in front of me were naked. I'm a bit of a comedienne I think. I saw a fat one in the front row and the thought of him naked just made me feel hysterical. But somehow I managed, we all did, we had to. On this particular occasion I got mad. Another platform speaker talked about Nottingham scabs. I was furious that he inferred that all the Nottinghamshire miners were scabs. I reminded everybody of the facts! Forcibly!

I still prefer to do administrative work for the group, but politically I've changed so much. I never miss a chance now to start a discussion wherever I am, in a bus queue, in a shop, anywhere I can get an audience, and I always manage to make the political points that are so important to me. I used to think Tony Benn was really way out. Now I wonder how far I can go before I fall off. I've become particularly interested in Northern Ireland, I'm reading up everything I can find about it. I just haven't got enough time to do all the things I want to do, to learn all the things I want to learn. When I compare my first speeches with those I make now for National Women Against Pit Closures I can't believe how far I've come, how simple my views were once.

I remember once during the strike going into a shop for some oranges. Liz Hollis, who was very involved in working for the Anti-Apartheid campaign, had told me about South Africa and the need to boycott the fruit from there. I asked the man in the shop if the oranges were from South Africa. When he said they were I said I wouldn't buy them and I told him why. I said I'd have Cape grapes instead. My geography was so bad that I thought Cape grapes came from Australia! He had to tell me that they were from South Africa as well. I've had so much education since the strike started and it goes on and on.

We all learnt, all of us, my husband Dave and my son Danny. He knows all about racism and sexism and ageism, all the 'isms'. When we tell him off he says we are being ageist! One time in the strike when I was away a lot and busy with the central group at home, Dave dragged me into the kitchen and showed me the stove. He said, "In case you have forgotten, this is the kitchen, this is the stove, and if perhaps some time in the future you would like to use it, this is how it works". He did the cooking, what had to be done; he did the cleaning, everything. My son Danny said to me one night, "Mum, have I got a one parent family now?"

Original Committee

Chairperson:	Ernie Daglish
Vice Chairperson:	Ida Hackett
Secretary:	Julie Wilkinson
Deputy Secretary:	Andy Miller

Final Committee

Chairperson:	Ida Hackett
Secretary:	Pam Oldfield
Treasurer:	Jackie Naylor

Members: Rita Abbott, Audrey Mulholland, Margaret Brown, Linda Hickling, Sharon Read, Liz Hollis

THE NUPE OFFICE

Without the organisational and administrative facilities provided by the National Union of Public Employees (NUPE) in a small office at their Nottingham headquarters, the work of the central group could not possibly have been as successful as it undoubtedly was. The office, situated in Sherwood Rise, Nottingham, served as the communication centre for all the support groups, both north and south of the county. It was acquired in May 1984 through the South Nottinghamshire Strike Centre. The men originally wanted it to be used for the south area only. But it was becoming increasingly necessary to have a base for the work being done by the central women's group, and it was felt that it was essential to cover the whole county. The women's group was already organising and collecting from the city stall. Publicity and fund-raising materials were needed, licences for collections had to be organised, and maintaining regular contact with each support group was of growing importance.

The office was given over to the women complete with all the facilities they needed, and at no charge whatsoever. Telephone, typewriter, stationery, postage, photocopying and printing facilities were all provided.

It was staffed mainly by the same women who organised the city stall, with others joining in during their holiday period or helping out for occasional days. There was a need for one or two women who could guarantee to be there each day, not only to involve those who came in on a casual basis but also to oversee and plan the work over a longer period. For the most part there were only about four or five 'regulars' and maintaining continuity was something of a problem throughout the dispute. The original intention of the women who set up the office was that it should be handed over to women from the mining groups. This proved to be impossible as most of these women were totally immersed in their work locally and because transport into the city was often very difficult. Although women came to help in

Notts Women's Support Groups.

4,000 MINERS ARE STRIKING IN NOTTS.!

4,000 MINERS HAVE STOOD FIRM AND WILL STAY RESOLUTE UNTIL WE HAVE WON !

FOR 7 MONTHS THESE MINERS AND THEIR FAMILIES HAVE SURVIVED WITHOUT UNION FUNDS

As you probably know, unlike areas where the strike is solid, here in Notts we have no access to union funds, and we are totally dependent on donations of food and funds from groups and individuals such as yourselves.

After weeks of strike, great hardship is being suffered by families - families are under threat of eviction from council homes, parents are unable to buy warm clothes and shoes for their children, central heating appliances have broken down and families can't afford to have them mended - as ever food is in short supply.

Women in Notts have taken on a great burden of the work - speaking at meetings, organising soup kitchens and distributing food parcels, the work is hard and tiring but we are determined to stand firm, and knowing that we have your help and support we have the strength to carry on and win!

WE NEED YOUR HELP NOW. MORE THAN EVER BEFORE !

Funds are desperately low. We are in acute need and would be very grateful of any support you can offer us.

Therefore we appeal to you and your organisation to send money and/or food, on as regular basis as possible. (Cheques payable to Nott's womens support groups, account no 2225-2258666, Trustee savings Bank, clearing code 77-22-28, or direct to the treasurer, or can be sent to our office at NUPE to be forwarded).

NUPE Offices, Sherwood Rise, Nottingham.
Tel. Nottingham 603522 - extension 29.

the office throughout the summer of 1984, and though they were always welcome, it was not always possible to involve them fully in what was going on. In the autumn of 1984 some of the original workers got jobs and had to leave. Around the same time some women who had come from Greenham Common became part of the regular rota; but on the whole help was erratic and no-one ever knew who would be able to come, and for how long. Liz Hollis from Nottingham got more and more involved and eventually became the mainstay of the office. Yvonne Woodhead from Blidworth was one woman from a pit support group who was also able to make a regular and continuing commitment to its work.

Help from regular NUPE staff was wonderfully generous, especially from members of APEX (Association of Professional and Executive Staff). Secretaries, telephonists and receptionists cheerfully coped with an extra burden of work about which they had not been consulted. After an initial reaction of surprise at the strange women doing masses of photocopying, and the additional load on the switchboard, an attitude of support and cooperation developed and the two groups of women became friends, working together in harmony.

The office provided a life-line not only between the groups in the pit villages, but also with the rest of the country and abroad. Arrangements for speaking tours and holidays were made from here, as were those for the donations of food and toys which came in from all over Europe. It was also a place where visitors could meet women involved in the strike, and where links could be arranged with the local groups. The office acted as a storehouse for all kinds of fund-raising materials and it was usual to find large dustbin liners full of donated clothing crowding the already over-full room.

Work was varied and covered everything needed by the women's support groups throughout the strike. Every week a mailing with minutes and agenda of meetings, notices and practical information was sent out. Links with trade unions were set up and maintained, parties and other entertainment functions were arranged; transport for adults and children to rallies and other events all over the country was co-ordinated, with lists of those requiring seats and planning of pick-up points throughout the county. At Christmas the huge number of presents flooding into the county caused massive organisational problems, as did the thousands of frozen chickens and turkeys. Some of the latter arrived late, causing panic that some families would go short, and presents were still arriving well into the new year. Ensuring equitable allocation was of major concern and sometimes impossible.

A major event was the 'Food for Victory' festival held on the Forest. The convoy of food and buses for carrying people from the Nottinghamshire pit villages came from Sheffield Trades Council on July 8th. The office had to organise the round trip to collect people who wanted to come to the festival from all over the county, to provide refreshments, entertainment, toilet facilities etc. for an expected 10,000 people, to let stalls and arrange games. The weather was beautiful and the festival was a highlight of the strike.

The relationship of the individual support groups with the office differed according to the degree of independence, but all groups depended on the office for communication.

Women who staffed the office:

Liz Hollis	Kierston Hunt	Yvonne Woodhead
Maggie Hood	Linda Deegan	Terri McDermott
Lynn Beaton	Chris Drake	Pam Oldfield
Doreen Gladwin	Tina Yemm	Margaret Davies
Ann Perry	Lesley Brown	Bronwyn

THE CITY STALL

All over Nottinghamshire small local groups were established to try to raise money for the dispute. In some towns, such as Newark and Mansfield, people tried unsuccessfully to get permits to collect. But despite the efforts of one or two pro-strike Labour councillors, neither these attempts, nor those to set up collecting stalls, were successful. However, in the centre of the City of Nottingham a stall was established and collectors were regularly out on the streets braving hostility, abuse and harassment from the public and the police. Not only was the stall very successful in raising money, but it also acted as an information service to counteract the adverse publicity and lies told about the strike.

From April 1984, between ten and three every Saturday, shoppers walking down the pedestrianised way from the square to the Broad Marsh Centre became used to seeing an extra stall, sheltering under the wall of St. Peter's Church. It was next to the Campaign for Nuclear Disarmament and Animal Rights pitches, both of which were long-established charity collecting points. The new stall was organised and staffed by a group of city-based feminists, women from various political parties and trade unions. They collected money and sold miners' lamps, mugs, badges, posters, tea towels, books and pamphlets, donating all the proceeds to the Nottinghamshire Women's Support Group.

As spring changed to summer and the strike went on, from autumn into winter and then spring again, the women enjoyed or endured the weather. At first sodden wet and shivering, they later became suntanned and freckled; then autumn winds swirled away their posters, papers and pamphlets, blew grit into their eyes and almost lifted their flimsy stall clean away. In winter everything turned soggy in the sleet and rain and they stamped about in boots and rubbed their bemittened fingers, almost too cold to rattle their collecting tins. Their noses turned red and dripped; they shivered as the news of the strike was bad and getting worse; and people hurried past, only the faithful having time for the stall and the miners.

Whatever the weather, whatever the prevailing mood of the shoppers, however they felt themselves, the women tried to be lively, good-humoured and optimistic about the outcome of the strike. Different women dealt with apathy or hostility in different ways. If Liz Hollis and Joan Blackmore were on duty, passers-by would be treated to a sort of double-act question and answer routine, deliberately loud enough to draw attention to the collecting boxes. It was both playfulness and bravado at the same time:

Liz: Hey, Joan, do you know who phoned me this morning?
Joan: No, I don't know who phoned you this morning. Who was it?
Liz: The Queen phoned me this morning.
Joan: Go on, she never did.
Liz: Oh yes, she did, and do you know what she told me?
Joan: The Queen told me to tell everyone in Nottingham today to give generously for the miners on strike. So come on everyone, the Queen says give for the miners, coal not dole.

★

Liz: Hey, Joan, do you know that in Annesley alsations are being put down because the miners on strike can't afford to buy dog food.
Joan: Does the Queen know that dogs are being put down because the miners have nothing to give them to eat?
Liz: Yes, the Queen knows that alsations are being put down and she wants people in Nottingham to give us money for dog food. So, come on, everyone, give for the miners' pets; dogs and hampsters and rabbits and budgies deserve to live.

★

Liz: Do you know, Joan, that Jesus is on the side of the miners on strike?
Joan: How do you know that Jesus is on the side of the miners on strike?
Liz: Well, he would be wouldn't he, being a carpenter and not being in a union. And he died for his principles. So he would be on the side of the miners on strike, now wouldn't he?

Such semi-serious, semi-comic turns were guaranteed to draw the

Raunds,
Northants.
Dec 1984

7.12.84

Dear Ladies

I have sent my small Xmas Contribution for the children... and Elsewhere contributing

The enclosed is for a couple of bottles of sherry and a couple of packets of crisps for the ladies of the committee to toast each other on Xmas Eve, or some time. To really congratulate each other; to have a "love-in"... and I wish that I could be there!?

God Bless You all

John Birdsall

The Nottingham City Stall

attention of passers-by. The conversation about the Queen and the pets brought marvellous contributions as people seemed to care more about animals than about people. On one particular Saturday, droves of people turned up loaded with tins of pet food and packets of budgie seed. The women were astonished but learned that this was the result of an interview on Radio Nottingham when a striker had explained the problems of finding enough money for pets' food when they had so little for themselves. Evidently this had touched the heart strings of a woman listener who had phoned in urging people to take food to the stall. The immediate problem for the women was one of transportation! Again, when Animal Rights collectors were doing particularly well and the miners' collection was low, they would transfer a tin from the miners' stall to their own, and return it when it was full, a nice gesture of solidarity at least with the animals.

The dialogue about Jesus brought down the wrath of an adjacent itinerant preacher. He loudly condemned Joan, Liz and Arthur Scargill to eternal damnation: "He told us that the Bible teaches us to look on our employer as a parent and obey him as one would a parent. Arthur Scargill, by going against this teaching, was the devil incarnate and we were his handmaidens in sin."

The stall was raising in the region of £60 to £150 every Saturday, so the women had to be careful to do nothing which would in any way hinder their work. They took turns at the stall, usually in fours, for an hour at a time. Sometimes women from pit support groups, especially Cotgrave, joined the regular workers. As a group they met regularly at the Narrow Boat pub in the city centre and sent delegates to the weekly meetings of the National Women's Support Group.

The site had been used for some time by various charity groups and since they had the support of the vicar, people had assumed that there would be no trouble. Certainly no-one in the past had bothered about applying for a licence to collect. However, appearance of the miners' stall was the signal for police investigation and harassment, so that all the stallholders on the site had then to apply to the City Council for permits. The Labour-controlled city council, unlike most of the others in the county, agreed to issue them. However, one very wet afternoon in October, two policemen appeared at the stall and told the women to pack up and leave. When they protested that they had been granted a permit, the police refused to accept this and threatened them with arrest. The women were in something of a dilemma, not being too sure whether or not the permit had actually been issued, and whether or not to stand their ground against two very aggressive and intimidating policemen. They were informed that "We are the law, and we are never wrong." It was pouring with rain and Liz Hollis had her umbrella up:

> When it accidentally touched one policeman on the shoulder he grabbed me roughly, accused me of assault and told us for the last time to pack up or be arrested. We didn't know what to do, we were watching the two policemen walking slowly down the street, radioing in to their base.

The police walked back in the women's direction, but to their great relief and surprise, they walked past, heads turned away, as if the stall did not exist. They must have found out that the permit had been issued after all. Thereafter, the precious piece of paper was always sellotaped to the stall.

But threats came not only from the police. Well-dressed middle-aged women were fond of trying to intimidate the collectors: "We have relatives in the police force", "I know a magistrate", "I will have you removed from the streets."

> Some people called us 'dirty scum' and we never really knew what to expect from members of the public who approached the stall. It was not threatening if they just walked by, but when they came up close, we felt apprehensive and vulnerable. People often came and asked lots of questions

The women never knew in advance why, and tried to be polite and appear relaxed, avoiding conflict and confrontation, knowing that the police were always just around the corner, waiting for the opportunity to move in to make arrests and have the stall banned. It was often necessary to bite back the sharp retort which came readily to their lips in response to insults or abuse.

Many sympathisers came to the stall not only to make a donation but to ask for reports on the strike activities in Nottinghamshire, so the women had to be up-to-date with what was happening. Often it was essential to counter adverse publicity and false propaganda. Sometimes they wished that they could be tougher. The older women in particular felt that they were too easily intimidated. They envied the younger ones who were bold and could parry the insults and deal more confidently with 'cranks'. Often when this was done particularly well, applause would come from their neighbours on the CND and Animal Rights stalls. If at any time there was only one woman on the miners' stall, someone from the other stalls would move over to give moral support, and once when two women were being threatened, unknown women who were passing by came to their assistance without being asked.

The worst abuse and threats usually coincided with media reports of picket activity: "One Saturday, collecting specially for Christmas, we stood cheek by jowl with Evening Post sellers shouting, 'Taxi driver killed, miners murder taxi driver.' " Needless to say, as well as bringing heightened attacks and abuse, these were also the occasions when collections dropped dramatically:

> We particularly hated being on the stall on Saturday afternoon after the pubs had closed. Drunks came specially to be objectionable to us as women, it had nothing to do with our collections for the miners. Their taunts were often obscene and they caused us great embarrassment. These were very bad times because we never knew how far they would go. We were always aware that the police would take any opportunity to move in on the stall. Once when a drunken man overturned the stall, scattering everything over the street, two old ladies came to help us restore order and retrieve out belongings.

Supporters varied tremendously and there were a surprising number of £10 and £20 donations. Many local political activists made a point of visiting the stall every week, and then it became a meeting place for activists to discuss the local and national progress of the strike. A special category of women who came were the elderly wives of retired miners. Often they spoke about the Spencer Union in Nottinghamshire, how their husbands had mistakenly joined it and how they, the women, had always said that it was wrong. They would reminisce about past times in the mining industry, giving glimpses of history to women who had no background of the pits. It gave the collectors pleasure to fix a 'Coal not Dole' sticker to a Jaeger suit!

People often brought clothes and food to the stall and it was difficult for the women to carry such things away, so they asked people to take things other than money to the office at the NUPE headquarters:

> One particular donation we particularly remember is the body warmer brought for a miner's child by an old man of eighty. He stopped for a chat and after hearing about the children's needs, he went home by bus and made a special journey back that afternoon to bring the woollen garment from his grandson.
>
> ★
>
> As the months went by, we noticed that some people regularly visited all three stalls, performing their weekly act of charity, dropping a ten pence piece into each of the three tins. We realised that we had, for such people, become a respectable charity. After the strike, people continued to come to the stall to give as an act of protest for the way in which the miners had been treated. They appeared shocked, saying that they never imagined the strike would end in such a way. They gave particularly to support the families of sacked and imprisoned miners. A great deal of indignation and bitterness was expressed about men found not guilty in the courts but still not reinstated at their pits.

The women of the stall, many of whom did not know each other before the strike, say how they have enjoyed and appreciated the experience of being in such a lovely 'mix' of women, of all ages, life-styles, political and social backgrounds, drawn together with one common purpose. "We are continuing to collect and we would do it all over again, only with more courage and better organisation next time." They continue to meet together as a Nottingham city group.

Women who staffed the City Stall

Liz Hollis	Julie Wilkinson	Wendy Wood
Lee Harrison	Tina Yemm	Trudy de Haney
Joan Blackmore	Mary Donelley	Moby Farrands
Margaret Davies	Jane Burrows	Eileen Hepple
Lyl Galloway	Mary Pope	Elizabeth Plant
Gail	Doreen Gladwin	Kierston Hunt

Terri McDermott, and women from pit support groups at Cotgrave, Gedling, Calverton, Ollerton and Hucknall & Linby.

ON THE PICKET LINES

When miners were out on strike in support of wage claims, as on most occasions in the past, the issues were clear and straightforward. But a strike on principle is a far more complex matter, needing analysis, discussion and debate. It reveals the quality and dedication of a man or woman and goes far deeper than financial gain. The striking miner in 1984 had often to explain and justify to his wife and children what he was doing. This was absolutely vital since the strike meant a shared commitment and shared hardship.

It was this understanding and commitment in the 1984/85 strike which led the women to the picket lines. Scarcely any of the women had been involved in such a way before. Few had experienced any kind of shared protest or demonstration for any reason or cause. Now picketing was an overt expression of women's support and also of their own deeply held loyalty to principles and beliefs. Women also wanted to disprove the widely held supposition that they would be supportive of men striking for money but not for principles.

As well as this demonstration of loyalty and of understanding the issues involved, there was another reason for their presence on the picket lines. They picketed in order to get closer to the dispute and also to express their hatred for the NCB's pit closure plans as near as possible to the place of direct confrontation, the pit gate. It was an opportunity to release hostility, resentment and frustration by shouting abuse and singing. They believed, at first anyway, that as women they might also be able to shame men still working to come out and join the strike.

At first too, the women thought that they would be treated with more tolerance and sympathy by the police than the men experienced. They soon realised their naivety in this respect. At first the police responded to them as mis-guided women who might be putting themselves in physical danger from the working men speeding to work in cars, or even from the good-natured jostling of the men on the picket line. But when women began to organise women-only pickets, the attitude and behaviour of the police changed almost immediately.

At Thoresby in early May, we came together by word of mouth from all over the county, called to a 'Tupperware party' at 9.30 pm. I went with Val Picker, Edna Colston, Judy Sim and Sylvia Barton from Newark. Many of the women were young and didn't seem to have come from the immediate area. Lines of police were drawn up to prevent us from getting across the road so as to be nearer to the cars driving to work. Word spread in their ranks that there were women from Greenham on the picket and this led to an immediate change in attitude. From being firm, but polite, expressing anxiety that we might be injured if we moved across the road, there were immediately threats of arrest and rough handling. That very early experience gave us a good idea of what to expect in the future. Later on the police began to deal with women as they did with men, herding us into areas where we could neither see nor be seen by the strike breakers, prohibiting the use of words like 'scab' or 'traitor', using 'snatch' tactics for arrests, holding women without making charges, and imposing bail conditions which restricted further involvement. It soon became obvious that any attempt to challenge police rules would be met with intimidation and violence.

Most women activists went at some time to the picket line at their own pit. Initially these visits were to bring food and drinks for the men, though women frequently stayed to talk. Often a woman would go to the picket line before starting a shift at the support group kitchen, going to the cash and carry for weekly supplies, going to her own place of work, or getting the children off to school. Sometimes fathers took their older daughters on the small, local pickets to encourage involvement and understanding. For a number of women, going to the picket line became as normal and routine as going to the local shop or the school gate. In addition there were pickets arranged by the Nottinghamshire Women's Support Group. At delegate meetings of the group it was not unusual for a woman to make a specific plea for support for a forthcoming women's picket at her own pit. She would explain the need for a boost to morale, that it did the men good to see and hear women on the line. It also raised the women's morale. You could be sure of meeting others from various groups and parts of the county. Not only that, but it was a chance to hear of individual

PICKETTING CODE

PICKETTING CODE

Devised by Notts. Miners Support Group, for women's pickets. 21st May '84

1. The point of a women's picket is to persuade non-strikers to support the official NUM action.

2. No woman should be made to feel that she ought to picket.

3. We will not picket with the intention of arrest. We will not block roads or deliberately obstruct them.

4. Every woman will be informed that she risks arrest, because the police are making the rules up as they go along. Every woman will also be informed of her rights.

5. Each group will have a steward, who will record the women's names, addresses, responsibilities for children, and anyone they want to be contacted in the event of their arrest. Stewards will be responsible for reporting any arrests to 26094 Berry Hill Control Room and to a member of NMSG, and for contacting relatives and making sure that children are provided for.

6. If anyone sees a woman being arrested, we recommend that the witness doesn't risk arrest by attempting to intervene, but takes careful note of the police officer's number and their behaviour, and takes a note of the name of the woman being arrested.

7. We will stick together in one large group as much as possible. At the end of every picket we will leave in one group, and not in ones and twos.

iners Support Group,

and group experiences and to plan response to special difficulties being encountered elsewhere. Any success, for example in dealing with welfare authorities, was related, and this gave encouragement to other women; any special shortage of a commodity might be alleviated by another group's generosity. There would be discussions of national and local news, of the numbers of men who were still out at the individual pits, and of who had gone in. It was always the prime intention of the women to make a political protest when on the picket line, but a secondary activity was a social one and much valuable information was exchanged.

Often men were given bail conditions which specifically stated that they were to take no further part in the strike in their own county. This was an obvious ploy to reduce the number of men able to picket. Some men withdrew to the kitchens and spent the rest of the strike working with the women there; others withdrew into their own homes, brooding and resentful. Women who had perhaps never before thought of going to the picket line went in lieu of a husband, brother or father, taking his place in an act of defiance.

In a special category are the women who picketed because they themselves were on strike, women such as Chris Brown who had been an office cleaner at Harworth Colliery and who was sacked for being on strike, though a trumped up reason was given for her dismissal. Women from pit canteens, such as Kath Black, were also on strike and took part in all the activities of the dispute.

Vickie Smails describes experiences over a whole week when she went to picket at the Orgreave coking plant.

> It was the week before the 'battle' of Orgreave in June. I wasn't a bit surprised to see the police violence which occurred then, they were getting ready the week when I was there. They just allowed their great alsation dogs off their leads and let them run about biting anybody who was near. I think they were practicing. That was frightening enough, let alone what happened in the following week.

Women from Welbeck describe how their first attempt to join a picket changed from an exciting, rather daring escapade, with lots of laughter and fun to, a frightening experience of arrest. They had decided to go on an early picket and planned to meet other women from the central group in the Fine Fare car park at Sutton-in-Ashfield, just a few miles away from Welbeck village.

> It was about quarter to four in the morning and we set off in a mini-bus and one car. We had our story ready in case we got stopped: we were going to Barnsley to find out how to run a kitchen for our kids. We piled in, all laughing and excited. Then about a quarter of a mile outside the village somebody said there was a police car behind and suddenly a police van overtook and got in front, we'd got an escort. They stopped us and told us to turn round and go home. We don't know to this day how we had the cheek to argue cos we were petrified. But we told them we had to find out how to feed our bairns and that we wouldn't dream of going picketing.

The police finally accepted their story and said they would be escorted to Barnsley.

> Now we were really in a mess. Nobody knew the way and anyway we hadn't got enough petrol to get to Barnsley. They said if we went to a picket line anywhere we'd be arrested. Well, we set off with the police in convoy. At Cuckney they decided we could go on our own, but pointed out that they'd got our numbers so we'd better not go picketing. Though we weren't far from Welbeck, we were completely lost. We were falling about with laughing when we saw a convoy of Yorshire pickets coming into Nottinghamshire. We tried to follow them, but they were going so fast we could hardly keep up. Eventually we got to Sherwood pit ready for the 5.30 am picket. We knew we might be arrested, but there were some Yorkshire women there so we got in with their group. We linked arms and started walking up and down, singing strike songs and shouting at the scabs. Then we decided to sit down in protest, all of us, in the road.

When Sandra Taylor was told by the Inspector to get back on the pavement or be arrested she decided she wasn't brave enough to be arrested; also, she did not want to jeopardise her fostering of children, so she did as she was told. Nine women from Barnsley and two from Nottinghamshire, Chris Gustard and Lyn Newburn, were carted off to the police vans and put into cells. It was the first time women in Nottinghamshire had been arrested on a picket line and they didn't know what to expect. Chris and Lyn describe the compound at Hucknall where they were taken:

> There were reporters swarming around. We thought we'd become famous all of a sudden. What we didn't know was that Ann Scargill had been arrested at Sutton, where we would have been if the cops hadn't stopped us. The attention was for her, not for us, after all. There were all sorts of rumours buzzing around and it was all very exciting. The police asked us a lot of questions they had no right to ask: Who did we vote for? Were we in a political party? What sort of financial position were we in? They took our photographs and they searched us, I don't know what for; and we had to hand over our jewellery, shoe laces and matches. What did they think we were going to do? Even though it was May, the cells were bitterly cold and we were really shivering. The food was just abominable, like pig swill.

> We were in court in October, it was my birthday. We got bound over, not to go picketing again or take part in anything in support of the strike. We just went on picketing, took no notice.

Toni Bennet of Bolsover was arrested on a picket line for a supposed breach of the peace, using threatening words and behaviour.

> I'd been singing actually. When they took my photograph for their records I took off my jersey and pushed out my chest. I'd got a 'T' shirt on with the slogan 'Balls to McGregor'. I wanted to make sure they got the message loud and clear.

In Bolsover it was the custom for the families of strikers to stand on the Wimpey estate near a bus stop, shouting to the men going in to work at the pit. One day a lot of police officers appeared from nowhere and one young policeman took the arm of an older women and said, "You look old enough to be my mother". She asked him how old he was and hearing that he was 24 she touched him on his arm in the same way and said, "Yes, but if you were my son you wouldn't be here.'' To her surprise he spun her round and said that that was just what he'd been waiting for, an assault. He dragged her away to the van cells, tearing her anorak in the process. Her distress was obvious; her daughter, with two small children screaming, was then arrested for trying to help her mother, followed by her son, who was knocked to the ground. Finally, it was the turn of her son-in-law; the whole family was arrested in one fell swoop. They were all released later with no charges having been brought against anyone.

Sharon Reed, active in the Ashfield Women's Group throughout the strike, relates the story of her arrest along with Lou Dove and Janet Gibson, also from the Ashfield group, and Nicky, a supporter from Cambridge.

> This was a women's picket, the biggest for a couple of months, and we'd organised it through the central group. We thought it was time we did Bentinck. I remember getting up for the quarter to six shift going in, it was September and beginning to get really cold in the morning. It was funny, the night before I'd had a bit of a tiff with my husband Brian and in the morning before I went out he said, ''Go and get yourself arrested.'' He didn't expect that I would!
>
> Well, we got there and the men started going in. This scab, he said to Lou, ''Why don't you get back where you belong, get off home and do your housework.'' Lou was sitting on the wall and she shouted back to him, ''I hope your next shit's a hedgehog.''
>
> Well, that was it. There were loads of coppers there and one grabbed Lou and dragged her off the wall. I wasn't letting that go so I rushed to the copper and protested that she'd done nothing to get arrested for. There was one particular one, he'd had his eye on me, thought I was the ringleader. They do that you know, keep looking, staring at you, it's meant to frighten you. He grabbed me and pulled both my arms up behind my back. He hurt me, bruised me under my bust, it was real rough handling. I was struggling. Then there were four of them on to me.
>
> Janet didn't like what was happening so she moved in and promptly

got arrested for assault. Nicky from Cambridge was arrested for shouting 'scab', so that was four of us. We got dragged into a room at the back of the pit. My arms were twisted behind my back and it was really painful. I told the coppers I wouldn't attack them but they said I was a very violent person. I'm not very big and there were four of them, what was I to do? My leather coat was hanging off my back. I was in a state but I wasn't frightened.

They took us to a van to get our photos. I told them they'd really hurt my arms. One policeman replied, "I'll hurt your legs as well if you don't look out." They were very interested in the central group, what we were doing, who was in it, the leaders. We didn't tell them a thing, but they kept on about it. After that they took us in a little van. They kept driving round in circles, waiting for women coppers to come. There's little sliding windows, like portholes, in the vans. I got to one and tried to pass my car keys to somebody. In the end I threw them to the women; they came running to get them but the police jumped on them, the keys I mean, and they wouldn't let me have them back.

When they began to interview me I told them, "Before we start I've got a complaint to make. It's against the law to take people's possessions away from them." The arresting officer made the copper give them back, so that was good for a start.

In Mansfield, at the police station, they put us in the 'drunk cells'. I reckon they did it on purpose because they were dirty and smelly, horrible. From there they took us to Hucknall. They wanted me to take my jewellery off but I wouldn't, I refused. We had to take off our shoes. I told them I wouldn't give them the satisfaction of me hanging myself, they could have them. I wasn't a bit frightened but Lou was shook up. I was all for refusing to have my finger prints taken. I knew my son Brendan was with his dad so I wasn't worried. But I decided to give in for Lou, because they'd have kept us in longer. We were kept from 6.30 in the morning to 4.30 in the afternoon. They rang Brian to tell him where I was.

At Hucknall the women coppers kept going on about the Greenham

Common women. They were really slandering them, saying filthy things about them. They said, "We hope you're not like those dirty women, they're disgusting." They kept on again about the organisation of the Notts central group; who ran it? what things did we do? where did we get our money from? They were digging deep that day to try to find out all about it, but they got nothing from us.

When our case finally came to court I was charged with besetting and with obstruction and insulting behaviour. Janet got charged with besetting and assault on a police officer, the other two with threatening behaviour and besetting. All serious charges. I was a bag of nerves in court. The prosecution tried to frighten us off, told us we'd have to pay police expenses of £2,000 if we lost. We wouldn't admit we were guilty, we wouldn't do a deal. He kept changing his mind, offering us all sorts to get us to agree, to buy us off. Liz Woodcraft, our solicitor, was relieved we wouldn't agree. It was worth going to court just to hear the police evidence. It made us laugh and we got into trouble for that. It was total imagination, the things the police said. They made it up. Twelve coppers all said the same thing, you could tell they'd got together. But my arresting officer, he must have missed all the talks and the agreement beforehand, cos he told it different. We had to laugh, you've never seen anything like it.

The magistrates here in Notts were much tougher than down in London. But on the first day we all got off on besetting. They found us all guilty on the other things, fined us £125 each and we got bound over for a year. If we got into trouble again, they threatened to bring us back to pay the £2,000 costs and take this case against us as well. They were threatening us if we went on in the strike, picketing. Between sittings, the police were all in the pub, laughing and giggling about the case, like a great big farce. They thought it was funny. I didn't. I felt lost, let down. I started thinking this was justice. Society was letting me down: I always thought the police were there to protect me. I never realised what they were really there for. I cried with relief but I felt let down.

Badges and 'T' shirts were produced in Nottinghamshire bearing little cartoons and slogans about these court cases, including one of a hedgehog and Lou's comment to the scab.

All the women who picketed tell graphic stories of abuse by the police. In Hucknall women were told, "If you need money for your kids, go on the streets, you'd make good whores. You're all cows, all of you." They have been pushed around and jostled when trying to picket peacefully. Miners going to work have exposed themselves in front of the women and have not been arrested. Police have lied in court in order to gain a conviction.

There are so many individual stories, too many to tell. However, the picket at Gedling in October is particularly interesting because it shows how police were by then unconcerned about criticism of their tactics or condemnation of their behaviour. I joined the picket at about 9.30 pm. There was a small men's picket and ours was about ten yards away, about thirty women. There were women from all over the county and a few from Derbyshire, all very good-humoured and enjoying being together. Two women from Sheffield Police Watch arrived and some Nottinghamshire members of the National Council for Civil Liberties. In addition there was a women's film crew from the London Film School. Nottinghamshire County Councillors Gordon Skinner and John Hepple were on the picket, as well as some BBC staff in a private capacity.

We started to sing. Ruth McIntosh and other women were warned about shouting 'scab' and Ruth was told not to hold up her placard. As the cars started to arrive, speeding down the lane, the police gave more warnings. They were all lined up facing us and it was impossible to see the miners going in or for them to see us. It was doubtful whether they could even hear the words of the songs, though presumably they knew what was being shouted. Suddenly a policeman lunged forwards to make an arrest. Before be could reach the woman he wanted, other women crowded around her to prevent her being taken.

"Take one, take all", we shouted. There was a reforming of police lines and by now experienced in police tactics we knew what was being planned. The police were preparing for a 'snatch', so it was essential to get the woman they wanted away from the picket line. Judy Sim gave the woman her coat to try to cause problems of identification and then a small knot of women tried to move away with her to get her to her car, and home. The police forced the group back and it was evident that reinforcements were being brought up to achieve the arrest. Photographs were being taken by the film crew and observers were recording everything that was happening. But despite all this, extra police were brought up, as well as women officers, dogs and cell vans. Finally three arrests were made. The woman originally marked out was suddenly surrounded, other women were knocked out of the way, the snatch was made and she was pulled and dragged down the road to the cells.

When Jackie Naylor was arrested it was assumed that she was a member of NCCL and she was questioned at length about her membership and her activities. The next day, reports on Nottingham Radio stated that the NCCL members had behaved in an inflammatory way, that it was their presence on the picket which had caused the trouble, and that one of them had been arrested. It cannot have been coincidental that the NCCL had recently announced that they would be setting up an enquiry into the policing of the strike and that they would publish a report. The statements put out about the picket at Gedling were obviously intended to damage the findings in advance by inferring that it was not impartial. Jackie describes her subsequent trial:

> There were two policemen giving evidence against me. The first said I was running about in the road, flaying my arms about. He said I was hysterical, out of control and inciting other women to cause a big disturbance. He said I was intimidating the men going to work. That was funny, really, because the police were all in line in front of us and we couldn't even see the men going to work, let alone intimidate them. What really happened was, I said 'scab' and the policemen told me I couldn't say that word and that if I did I'd be arrested. I told him it was a proper word, in the dictionary, and I would go on saying it if I wanted. I said it again and I was arrested. I didn't struggle and I didn't get hurt. But when I was in the van and I refused to let them take my photo, they pinned my arms behind my back and pulled my hair, yanked my head back to force me to be photographed. In court the other policeman told the truth and so there was a real problem. I was fined in the end, £140, but I wouldn't pay it and I appealed. I got off then and I'm taking a case against the police. They lied through their back teeth.

Jackie had been charged with insulting language and behaviour likely to cause a breach of the peace, and also with the very serious charge of besetting. In Mansfield Police Station she had been told that her solicitor was advising her to allow her finger prints to be taken, when she had refused. This was later found to be a lie. Jackie's experience leaves her with no faith in the judiciary system, but with a firm conviction that people have to stand up for their rights, whatever the possible consequences.

Brenda Greenwood, who spent a week in the Remand Centre at Risley, later defended herself in the magistrate's court in Nottingham and had her charges dismissed. She too stood up against the police for her democratic rights. She did not give way one inch through the anxious weeks of the remand period and three appearances in court.

Events did not always turn out so dramatically. At Bevercotes, we had one wonderful picket in the snow and even turned back two cars. Maybe the men were not too keen to go to work that night, but we rejoiced anyway. We had a free hand with only two policemen on duty. We danced around in the road, waved our umbrellas, shouted whatever we liked, sang our songs and had our photographs taken by Brenda Prince. By this time, in February, we knew that there was not much chance of stopping men from working, but we tried anyway. One ploy was to arrive in one large group for a picket, usually around ten o'clock. There would often be only a token police presence but when so many women arrived calls would be made for reinforcements. In response to the radio messages, vanloads of police would arrive. not exactly pleased at being called out late at night. As soon as they were out of the van, overcoats on, helmets straight, women would say a polite but gleeful 'good-night' and go home to their beds.

The traditional song of Nottingham took on new words:

> Maggie Thatcher, Maggie Thatcher, riding through the glen,
> Maggie Thatcher, Maggie Thatcher, with her band of men,
> Takes from the poor, gives to the rich,
> Robbin' bitch, robbin' bitch, robbin' bitch.

and

> Build a bonfire, build a bonfire, put the Tories on the top,
> Put McGregor in the middle and burn the bloody lot.

''Maggie Thatcher has one, Ian McGregor is one'' was dropped from the repertoire as women became more aware of its sexist implications.

All remaining vestiges of a good relationship with the police quickly evaporated as women were offered handfuls of grass for their children to eat and insulted in other ways. On one occasion a woman who had tried to put the miners' case was told: ''All miners on strike should be put underground, and the women with them and the pits filled in.''

Women on the picket lines were a study in courage and audacity. Eyes sparkling, fists waving, they shouted, sang and danced, arms linked in comradeship. Their excitement was contagious, especially to women who came from outside Nottinghamshire to pledge their support and experience a picket line. They picketed in the summer, in dresses and 'T' shirts. They picketed in winter in boots, top coats and woolly hats. Women of all ages and backgrounds told working miners in Nottinghamshire where to shove their national ballot.

OUTSIDE NOTTINGHAMSHIRE

One of the biggest challenges faced by women from the support groups was that of travelling outside Nottinghamshire to raise support. Women travelled all over the country and sometimes abroad, to present the miners' case and ask for financial and political help. They spoke to trade union branches, political organisations and women's groups, to students' unions, church organisations and many others. Most had not travelled far from their own territory before, and had never even addressed an audience of friends, let alone absolute strangers. They often suffered sleepless nights or days of anguish before a meeting, never sure of where they would be staying, always worried about their ability to answer questions.

Jackie Naylor and Maureen Gill will never forget their very first meeting in May, to talk to a group of delegates from different areas of AUEW TASS in Rugeley. They were real novices, and terribly nervous. Jackie remembers that the business part of the meeting seemed to go on for ages:

> We sat in the boardroom at the side, waiting for them to finish, but dreading when they did and it would be our turn. Maureen said I was to be first, we were smoking one after another. The longer we sat the worse we felt. Then I lit another cigarette and didn't notice that I'd dropped the head of the match onto my notes in my lap. I didn't know what was happening until my notes were in flames and Maureen was flapping her hands to put the fire out. I had to speak first, and without my notes. The people were grand and said they hadn't made any fuss over the fire, though they had seen what was happening, because they didn't want to make me more nervous.

Women describe headaches and stomach aches from the tension and worry. Julie Hunter from the central group went to London in June 1984, to a social at Finsbury Park and then a meeting the next morning:

> After the social I got to bed and suddenly remembered I'd left my speech in the car and the driver was staying elsewhere. We were not meeting till 10.30 and the meeting was for 11.00, so panic set in. I sat up till 3.30 am rewriting my speech and fell asleep still writing, woke up at 6.15 and wrote 27 more pages. By 7.30 am I was washed, dressed and feeling very sick. I couldn't eat or talk and sat watching the clock slowly tick the hours away until I was 'ready' to go. I wished I was somewhere else, preferably Mars! I bought a cheese cob on the station waiting for the train to come in from Nottingham, to meet my support for the meeting. I hoped I would catch salmonella but I didn't. In the taxi, I hoped we might get lost, but to no avail. When we got to the meeting I laid out my badges, stickers, pen, notebook and 93 pages of speech. I told Lesley I couldn't do it. But I did. I talked solidly for 12 minutes with very few pauses. I hardly even looked at my 93 pages of speech until at last I thanked them for inviting us. I was elated at completing the impossible.

Hundreds of stories such as this one, each with its unique individual experience, were recounted when women returned to their families after a meeting or a campaign. They would return joyfully with cheques and contacts for further support, well pleased with their efforts. But they frequently returned to urgent needs and problems at home. When they had to go away again they often took these problems with them, hating to leave behind children and husbands who were missing their care.

Ann Burrell, who had begun by making little drawings and verses which she attached to the daily menu in the strike kitchen, progressed to reading her poetry from a West End stage.

> I went to the Piccadilly Theatre. There was an audience of 2,000 people and Julie Walters, Charlotte Cornwall, Anna Cartaret and Ann Mitchell were on stage. Just imagine me, a little housewife from the backwater of Warsop, reading my poems! I came out feeling ceiling high, absolutely glowing. I thought I'd be out of my league, but I wasn't. Why couldn't I do it before? I couldn't describe my emotions before, it was as if I was tongue-tied. Then with poems, it was suddenly different, it was a sort of release for my

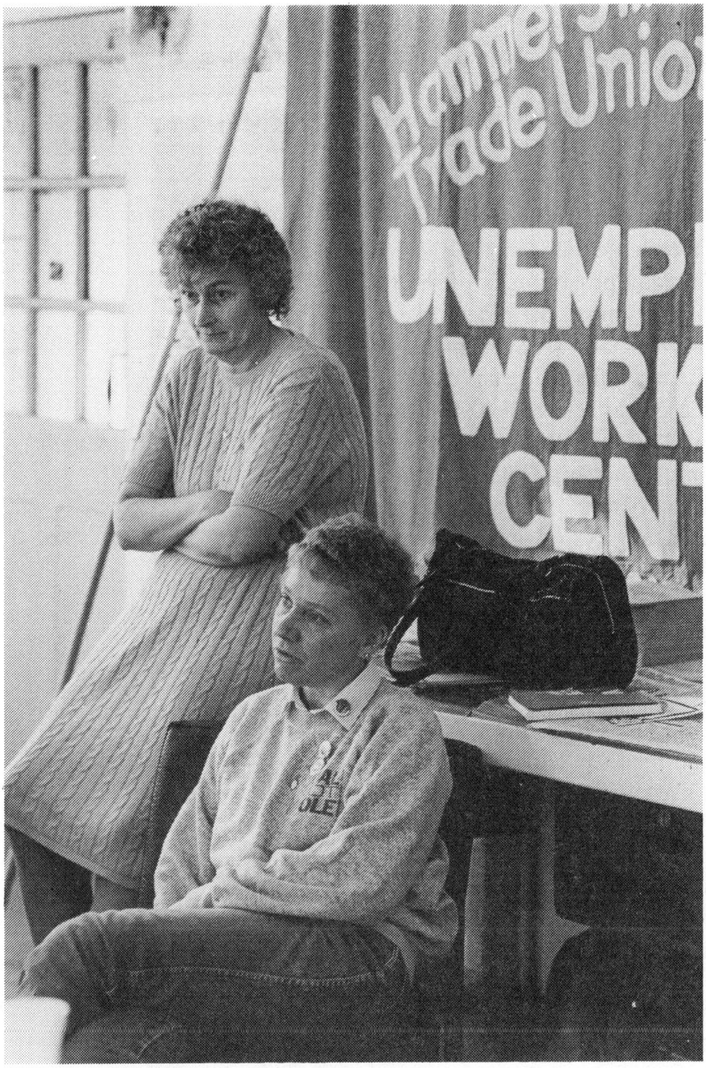

Brenda Prince/Format

Speaking at Hammersmith and Fulham Unemployed Workers' Centre: Marie Price (Ashfield) and Jackie Naylor (Mansfield)

feelings. I want to go to Northern College or take a pre-degree course to study Communication Skills at college in Mansfield.

THE DAY THAT GRANNY GOT NICKED

by Ann Burrell

On Grandad's knee we like to sit,
To hear the tales of the fight for the pit.
Grandad proudly tells of the time
Granny joined the picket line.
Not normally renown for her grit,
Granny refused to stay at home and sit.
She gathered together all her strength,
(well, as much as you can expect when you're five foot in length!)
And by her principles she stood and fought,
Which led to an appearance in the court.
Grandad says, it was the funniest sight he ever saw,
The day poor old Gran fell foul of the law.
There Gran stood trying to have her say,
But the bobbies on duty didn't see it her way.
A young constable told Gran she'd have to be quiet,
Or she'd be charged with inciting a riot!
Gran faced him with a wily glint in her eye,
And then she let out her battle cry.
It's a pity he didn't hear her shout,
He might have had chance to avoid the clout!
Again with her handbag she took aim,
But missed her target, what a shame!
Over to Gran a red-faced sergeant trotted,
His face all twisted, as if he were being garrotted!
It didn't take three chances or a prize to guess
That Granny had landed herself in a right mess!
Granny, God bless her, had refused to be licked,
So that was the day that Granny got nicked.
Although it happened many years ago,
I'm glad Grandad saw fit to let us know.
I love my Gran, she's one of the best,
What would you do if threatened with arrest?

But looking at Granny now, so wrinkled and old,
Who'd ever have thought she'd been so bold!

Sometimes visits to other places brought the dispute into perspective as part of a wider struggle. Elsie Henderson from Welbeck went to Derry to organise a women's support group and saw for herself how police trained in Northern Ireland had brought to the coalfields their techniques of control and intimidation. Her diary of the time she spent amongst trade union activists describes her feelings as she saw conditions there and heard from people of their own suffering and grief.

Wherever I went, everyone made me welcome. I travelled to Londonderry and when I arrived there were two bomb scares. One was detonated as we walked past a side street. I froze to the spot. I was amazed that no-one was hurt. I heard in the evening that the other one had gone off later.

On the Sunday Dermot took me to the march in Derry, in remembrance of Bloody Sunday thirteen years ago when the paras killed fourteen people. Six of the dead were only about seventeen years old. There were about three thousand people marching with banners, each group accompanied by a flute band. The relatives of the people killed all carried wreaths and placed them on a monument in the centre of where the incident happened. There were American Indians, some Navaho and others who sung a death chant. It was ever so eerie.

We've been to the older part of Derry, there is a great big wall goes right around the town. I've never seen so many factory buildings standing empty and going to waste. The unemployment is terrible and yet they want to give us support.

When we went over the border we got through customs alright but coming back the police wanted to know who we were, where we were going, where we'd been, did we have anything in the boot. They checked tyres, lights, tax disc, and were very interested in all my badges and Theresa's clipboard. But eventually they gave Dermot his driving licence back and we continued back to Derry.

Maureen Gill from Mansfield speaking in the U.S.A

Every time I see a policeman I get out of the way, it makes me nervous, even though they are part of the scenery; I don't like their guns and rifles. I wish you could see the police station itself, it's just like Stalag 19. I'm getting used to it a bit more, but I still get a twist in my stomach when I see the police around.

Maureen Gill was chairperson of Mansfield Ladybrook Support Group and was particularly active as a speaker, not only for her own group, but also for the central group in Nottinghamshire.

I was so used to being asked by Ida Hackett to go out all over the country to speak to trade unionists and Labour Party members that when she asked me on this particular occasion I automatically said yes. I had already been on hundreds of speaking missions so one more was no special event, or so I thought. When I casually asked where this one was and she told me it was America I couldn't believe it. It took me a week to stop shaking.

But what a marvellous month I had, from January to February 1985. It was hard, gruelling work, travelling long distances in Michigan and Ohio, all in the Great Lakes area. I was with two Notts miners, Mick McGinty from Ollerton and Terry Martin from Cotgrave, and we spoke to audiences at open meetings, at trade union meetings and Council meetings. We did an hour's TV interview for their Channel 4 and I did an afternoon women's programme. There were radio programmes which were sometimes interviews, sometimes question and answer sessions. It was a packed round of meetings, explaining what was happening in the mining communities.
The audience reaction was amazingly different from Britain. I found it a bit disconcerting at first to be interrupted with shouts of,
"Right on lady, you're handling yourself well honey"
"I'm listening doll, you's telling it straight to us, hit us with it baby"
The men were treated in exactly the same way.
"Right on man", etc.
This was the norm for all the meetings we did.
I learned a great deal about workers in America. In Detroit where steel was privatised, workers thought that private owners would take better care of them. It was the opposite and I saw workers

> living in absolute squalor. There is no health service so it's 'pay as you go'.
>
> We went on a picket line whilst we were in Toledo to support the A.F Parts strikers. It was very moving and significant for us, we were proud to be there, showing international solidarity as they were showing it with us. We were videod by armed security guards. We gave back some of the money we'd raised for the A.F Parts strikers.
>
> Friends of the NUM came to meet us from Canada when we were in Oakland Hills; they brought money and a TV company to film the handing over of the financial support. We watched a video of *Harland County,* about miners on strike in the past in America. We attended a Council meeting in Detroit City Hall at which a resolution of support for the British miners was passed; and we were given the Freedom of Detroit. Despite the worst weather I have ever known, with howling winds from the Arctic Circle, this was an experience of a lifetime. We raised a great deal of money, we met many wonderful people and everyone was so hospitable.

Trepidation about the kinds of people they would meet, the places they would stay, compounded their anxieties about presenting a good account of the strike. One day I sat with 'Nettie' Bone and Joan Preston in the support group kitchen whilst the ham and leek stew frothed and bubbled on the stove, talking about their stay in London with a group of women there. They described their horror of leaving Bilsthorpe and going into unknown territory, where the women they were to meet included prostitutes and lesbians, and where some of the women lived in squats. They found conditions very different from their own homes.

> We were like fish out of water. We couldn't believe that people could live in such a muddle; things all over the place and nothing belonged to each woman, they shared everything. What were they doing supporting us when they didn't have two pennies to rub together for themselves?

But differences were soon forgotten in the shared commitment to the miners' struggle and gradually women grew to understand and respect each other's different attitudes and ways of life.

> They were out at meetings collecting for us every night and on the streets collecting every day. They were marvellous women and we really liked them when we knew them. We forgot how different we were because we were really the same in lots of ways. They showed how good people can be, unselfish even when they haven't got anything for themselves. It was an eye-opener. We like the way we live, and we wouldn't like to have their way of life, but we accept them now and understand them a bit better. They learned about us and it is good to feel that women are linked together.

WOMEN'S PERSONAL STORIES

The following personal stories were not specially selected from among those of other women. I wrote them because I knew the women and something of their work and their experiences, because they volunteered to talk to me or because they were recommended by other women. I could easily have found two hundred and more women whose stories were equally worthy of inclusion.

Each story is unique, yet women all over the Nottinghamshire coalfield will recognise in the accounts many things which they did or felt during the dispute. So, not only do Pam, Brenda and the rest speak for themselves, each in her own unique way, but they also speak for the others. They will bring back memories of courage and fear, elation and depression, hope and despair, confidence and helplessness, anger and joy, gratitude and resentment, enthusiasm and exhaustion, all those extremes of emotion which they experienced together. They will recall for other women the pride and strength which they felt in the very act of working together for their common cause.

I had hoped to include in these personal accounts one from a Nottinghamshire woman who had opposed her husband's refusal to join the strike. No-one knows how many women fell into this category. A great deal of coverage was given by the media to women who supported their menfolk in demanding their 'democratic right' to break the strike but no-one heard of the women who were in conflict with husbands, fathers, brothers and sons, wanting them to strike.

I knew of one such women and tried to encourage her, through the wife of a striker, to tell her story. However, she felt that this would be the final straw for a marriage almost at breaking point. She would have given all her spare time to working with the women of the strike and realises now what she has missed in terms of relationships and personal growth. She knows, above all, that she can never feel the personal pride in sacrificing with others for the general good and having a partner who was in the forefront of that bitter struggle.

PAM ELLIOTT

Age 44, suspended from duty after Henry Richardson was sacked from his position as Area General Secretary on 14.1.85. Pam was finally sacked herself on 22.5.85

Before becoming personal secretary to Henry Richardson, I worked for Joe Whelan, former Nottinghamshire Area General Secretary, then Ray Chadburn, who was acting General Secretary for a few months. I have worked at Berry Hill, headquarters of the Nottinghamshire Area of the National Union of Mineworkers, for seven years. My husband was previously a miner at Rufford Colliery. Joe Whelan often talked to me, encouraging me to read *The Ragged Trousered Philanthropist* and *Germinale*. He was a great influence on my ideas of politics and of humanitarianism. When the strike started I believed that the fight was right and I wanted to be involved in helping to alleviate hardship wherever and whenever possible.

Men started coming into the office and at first Henry Richardson gave them money out of his own pocket. We did not make a plan of what we should do, it grew naturally out of seeing the growing hardship of the strikers. Of the people here, about half were sympathetic to the hardship but only about a third to the cause. The Hardship Fund which was set up at Berry Hill included the salaries of Henry Richardson and Ray Chadburn. A National Union directive had said that all area officials should give up their salaries to the Nottinghamshire Hardship Fund, but other officials disregarded the directive.

The fund also received donations from individuals, from the Nottinghamshire Women's Support Group and from West Germany via the Ollerton Strike Fund. Mary Lane, an old age pensioner, donated £10 every week in addition to food parcels and clothing. Miners out collecting sent money. Every week staff at Berry Hill got together an enormous food parcel. At first this went to individual groups, but then it was given to Ida Hackett to distribute to individuals with special needs.

As the strike progressed, there was never enough to meet the increasing needs. Our income fluctuated, but the demands were always heavy. No-one who came to me was ever questioned or asked for proof of hardship. Everyone was taken on trust. Help was given for children's shoes, for clothes, for chain saws and for logs. We bought permits for cutting timber in the forests, and anthracite for those who could burn nothing else in their heating sustems. We gave out 'holding payments' on gas and electricity bills to prevent supplies from being cut off. We also tried to make small donations for special events, such as buying flowers for the mother of a new baby.

Several weeks before Christmas we put a special collecting box in Berry Hill. Many of the staff contributed, some willingly, but others because the were shamed into it. Five women who used to put aside £1 a week towards a meal out together gave this money instead to the children's fund. They also bought an occasional bottle of whisky to give pickets a drink in the bitterly cold winter weather. When the family of an evicted miner was allocated a holiday in Devon, I stood over people to get contributions to give the family spending money. We raised £70 then, but generally it was always the same few who were generous.

One criticism used as an extra reason to sack me was that I worked too many hours each day for the strikers. During the strike I worked from 8.00 in the morning to 8.00 at night with no lunch break; I came in on Saturday morning and often on Sunday, to make sure that I never neglected my official work.

It might have been better if the Hardship Fund had been organised outside the Berry Hill office, but the office was a place where the strikers could come for a cup of tea and a chat. They used to come in earlier and earlier in the morning; so I used to come earlier and earlier to try and get my work done; then they'd come just on the off-chance I'd be there! I'd get nothing done, but it was important to see them. I shall always remember the strikers. They were so warm, so friendly, so good-humoured. And they were so loyal to the strike and to their union. When I asked them what motivated them to come out, they'd say that they believed in what they were fighting for, but, most of all, they would not cross a picket line. What impressed me most was the number of lads who joined the strike who had had no union involvement whatsoever.

I would do it all over again, but I would start earlier next time. I wish I could have gone on all the marches, and I wanted to go picketing and on demonstrations; but Henry thought this would have been too provocative. It would have put my job at risk, and then I could not have done the work for the strikers. After I was suspended I had no access to Berry Hill so I continued the work for the Hardship Fund from home, then from an office we managed to get in Wood Street, Mansfield.

Something that will live with me always was the day the strike ended, lads who had endured terrible hardship for twelve months without a murmur held me in their arms and cried.

Throughout the Nottinghamshire coalfield, the name of Pam Elliott is mentioned with the greatest regard for all her hard work and her concern for everyone during the dispute. Both men and women speak of her with love and affection and with great appreciation.

BRENDA GREENWOOD

Age 41 years, married to Keith, aged 45 years, miner at Bevercotes Colliery for 12 years. One son, Clive, a joiner, one daughter, Heather, aged 16.

We moved to Nottinghamshire for a job for Keith, where he could get the highest wage. We have no history of mining in our families. When the strike started and the vote here went for men to continue working, Keith did that, he thought it was democratic. He had had no experience of being in a union before he came to the pit and we were not political. What changed us was a friend who had worked with Keith for years at the pit. He kept coming to see us night after night after night, telling us about trade union history, showing us how unions had fought for better things for the working class. He showed us what the strike was for, and that decided us. Keith had not been happy, deep down, about going through a picket line and when I said to him, "If you feel that you must come out, then do it and I will back you to the hilt", he just looked at me and said, "That was all I was waiting for." From then on we never looked back. We were very green, but we learnt as we went along. We learnt about the struggles of other people; we had never thought about it before, it didn't seem to have anything to do with us.

About four weeks into the strike, when Keith had been out for one week, I went up to Ollerton, where the women had opened a kitchen in the Welfare annex. I went to the sink and for a couple of weeks I just washed pots, I've never washed so many in my life. Then some of us women decided to go down to the picket line. The police were there and said we could not stand on the corner of the pit lane. They had all the pickets herded into the pub yard, away from the men going into work. I stood on the corner by myself and was told to move. I could see no reason why, so I refused. I argued with the police and I didn't move. When we got back to the kitchen the women decided we had to have a committee and I was elected chairperson, for standing up to the police. I challenged the police because I knew I was in the right, no way could they say I was breaking the law. We were naive in the beginning, we soon learned that the police could do anything they liked.

At Ollerton we were first off the ground, with the Blidworth group, and we took the lead. When other areas saw what we were doing they wanted help to form their own groups. We started holding meetings to show them how to go on. Like many other women, I went out speaking all over the country. I was hardly ever home for days on end. My children used to make jokes about making an appointment to see me. What I was doing was so consuming, I felt so strongly, I just had to do it. I never had time to think I was losing a year of my children's lives and they were losing something I should have been giving them. I wrote a lot of letters during the strike. Keith says that when I write a letter I write a book. I had so much to say and I always tried to find exactly the right words. All the time I was out speaking and meeting people I was getting more knowledge, learning about the working class struggle. Everyone I met was able to tell me more and I spoke to them from my own experience.

In November I was arrested. I thought that it was time that we joined the men on the picket line again. We had all been so busy that we had neglected to support them, and it was a very bad time anyway. I went down to the pit corner again, as I had done early in the strike, and stood alone, opposite the men on the picket. I wanted to shout to those going in to work, "Look to your conscience, you are working for blood money, men have died to protect your job." Again I was asked to move and I refused. I told the police that I was demonstrating about an issue I had very strong feelings about and it was my democratic right to do this. I was arrested, but later released, having been warned not to return to the same place.

Next day I went, alone, to the same spot, and shouted to the men going in to work. I was arrested again and this time I was taken to Risley Remand Centre. I'd never even heard of Risley before I was arrested. I wasn't a bit frightened but I was concerned about my family, how it would affect them. I never wavered for a minute, other people wavered for me. Keith was frightened for me and so were a lot of women. They did not think that what I was doing would serve any useful purpose. But I had to go through with it, because I knew it was right. No way was I going to let the police take away my rights. I was never frightened. I knew I shouldn't be in Risley, but I knew I was in the right. I knew that I had to keep my dignity and keep my emotions under control whilst I was in there, not to let it get to me. I buried myself in books, and tried to forget where I was and what was happening to me. I got a lot of letters of support from outside. They were opened and read. It annoyed those in charge of me to see these letters, that gave me a lot of satisfaction, and the messages helped me get through the week I was in there. I was locked up for 22 hours on end, just let out for fifteen minutes for each meal.

I was surprised how good the other women were to me. When I first went in, nobody told me about my rights and in the morning I had to clean the cell and the landings. Then I was told to clean the toilets and the bathroom. One of the girls came up and said that I shouldn't be doing that. They did the work because they knew that they would be sentenced and sent down, so it was better to do what the warders wanted. But they felt that I was different and I shouldn't do it. The next morning, when I refused to clean the toilets, I was told that I would be locked up continually from then on.

The other women asked me about the strike and were very interested and supportive. They talked to me about themselves. They felt that society had put them inside; everywhere they saw huge glossy adverts telling them to buy videos, nice clothes, televisions, and they've got no money. So they turn to crime. The young woman in my cell was only seventeen, a drug distributor. She talked to me a lot. What do you do when you have no money? You go in and out of prison, with no dignity and no pride, you get what you can, take life by the scruff

I heard about your case at a Miners' Support rally at Wellingborough on Saturday.

I cannot express how strongly I admire your courage. We all like to think that when we come to "the test", that we will be firm in our convictions and refuse to submit to things which we know are wrong and unjust. You have passed that test and you are a fine example to us all.

I feel sure that you will have lots of support from family and friends when you have served your sentence. I hope that the thought of all the friends and strangers who are praying for you and wishing you well will help to relieve the gloom

Kindest regards
and very Best Wishes

Dudley, Tyne and Wear.
June 1984

of the neck, there's no other choice. They helped me to understand what society can do to people. It was another experience and I learned from it. I tried to keep my dignity. The warders were not used to that, nor to people speaking to them as I did. They didn't like it and they made snide remarks so that I should overhear, about the strike, about Arthur Scargill. They were against the strike. But if they tried to provoke me they failed, I never let it touch me. The worst thing was the noise on the lower floor at night. I think the women there should have been in a psychiatric hospital, not a jail. They made the most awful din, every night, banging on the doors and the grilles on the windows, it was really frightening.

My daughter Heather was very upset and all she could say was, "Why have you done this to me?" I tried to explain, but I don't think she ever understood, and this put me even further apart from her. Now, after the strike, I want to spend a lot of time with her again. I hope she will understand one day. As for Keith, we came even closer together than before. He was very proud of me and of all we did here at Ollerton. All the men were. In the beginning the men were running round in circles whilst the women were forging ahead. There was some jealousy at first but when they saw that it was the women who were holding things together, and it was the women who kept the strike going, then they had to respect us. Men thought they knew the women they'd lived with for years. Miners have always been thought of as male chauvinists, thinking women's place is in the house, washing pots and changing nappies, kept there by the 'boss'. Women turned that on its head, the men did those things while the women were keeping things going outside.

When I knew I had to face charges of obstruction, I decided that I wanted to defend myself, at least the first time I had to go to court. I had a friend, Martin Walker, to help me to prepare my case and tell me all about the proceedings. I had never been in a court before so I didn't know what to expect. I felt that none of the solicitors and barristers had been able to say anything political when they were defending miners and I thought that as a novice, I might have more leeway to get away with it. I wanted to go as far as I could in making a political statement about the attacks on our democratic rights, about the police, the courts and the government. It didn't seem relevant to me whether I won or lost. I knew I'd got a good case and I was prepared for

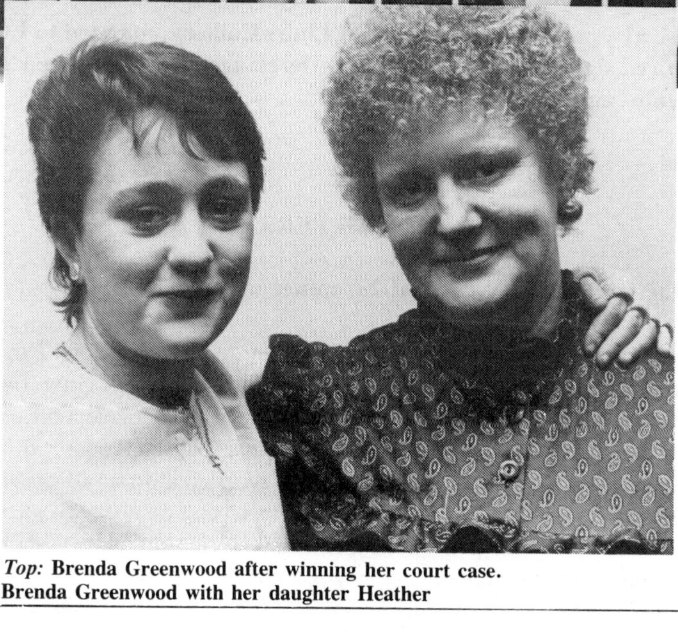

Top: **Brenda Greenwood after winning her court case.**
Brenda Greenwood with her daughter Heather

whatever was the outcome.

For the first fifteen minutes I was nervous and my knees shook, but after that I felt fine. The fear just went. If you can stay calm and dignified it shakes people who think they know better than you do. I just behaved like I would anywhere. If I'd had any doubt that I was in the right, I might have worried, but I hadn't, not at all. I started on my speech and I said a lot before the chairman of the magistrates got angry and tried to stop me. But I managed to finish off what I had to say. All my friends in the court thought I'd blown it, they said I was doing so well until the magistrate got angry, and they were worried for me. They didn't seem to understand that it didn't matter to me what the verdict was, I'd said what I wanted to say. I won. Three times altogether I was in court, once on my own and twice with solicitors, but every time I won. There was never enough evidence to convict me. I won and I proved that I had been right.

DOROTHY (DOT) HICKLING

Age 51 years, canteen worker at Linby Colliery, married to Les, retired signalman; four children: Derek aged 30, Daryl aged 28, Denise aged 25, Wayne aged 21.

and

DENISE PERRY

Married to Malcolm, aged 28, miner at Hucknall.

Dot: I've been a canteen worker at Linby for ten years and I'm in the NUM. There's nobody in my family from mining, I come from Burton in Derbyshire. I came out on strike when they made it official. Linda and Daryl and Malcolm, that's Denise's husband, came down here and we all talked about it. Les was worried at first, about what we'd have to live on. We only get £37 pension and we need my wages, I get about £78. We get Wayne's board, that's extra, but he was only an apprentice bricklayer.

Denise: Malc was one of the first to come out, he wanted the whole pit out. People think we wanted to come out on strike; nobody wants to strike, but you have to, for the principle.

Dot: We all went up to our Denise's house, it started from then really. We got the Labour Rooms and I just used to go up there and help Julie, cooking and washing up. It got so hectic, there were too many people running it so me and Carol Roberts became the supervisors, we ran the kitchen. I used to go picketing when I could. I went on women's pickets to Hucknall and to Linby. It was great, women from all over, all singing. Rita's husband was taking photos and when the cars came we shouted 'scab' and all sorts. One time it was the Hampshire police, they let us do what we liked. We had a real good time; it kept us going and the picture was in *The Morning Star.* We wanted to go to Ollerton but we couldn't get a car. The men were going to the power station so they had all the cars. I never thought I'd ever go picketing, never in my life. Les was a bit frightened, there'd been lots of arrests. He kept away from us.

Denise: But he was for the strike, my dad was. He'd do anything in the kitchen, or he'd help anybody, take a food parcel to their house.

Dot: We had a good time on the picket at Christmas. I'll never forget it. We had a great big cake, which somebody sent, and sandwiches and pork pies. There we all were, sitting on the picket outside the pit. There's always a fuddle down the pit at Christmas, we had ours on top; party hats, Santa Claus, beer, wine, we had the lot and they even had a mock arrest, it was really funny. It was good picketing, yes, but I didn't go out speaking, I'm too shy. I'd dry up if I got on a platform, I wouldn't know what to say.

Denise: Well, you'd got a full-time job in the kitchen, every day from 8 o'clock till 3.00, that was enough.

Dot: I went to some other kitchens, to Blidworth, then to Yorkshire and I went to see Cortonwood where it all started. We all went together on marches and rallies, I really enjoyed them. I went on the women's march in London. I wasn't political before the strike. I always used to go and vote Labour, but we never used to talk about it before. Now we talk about it all the time. I've learnt a lot. We went to Ollerton where David Jones was killed, for the memorial, and I heard Dennis Skinner. I've heard Peter Heathfield, Henry Richardson and I've heard Arthur. When the weather got bad we had to go to the Hardship Fund at Linby. Everybody was in the same boat. We didn't go for food parcels at the beginning, Wayne helped us a bit, then we got hard up with gas and electricity. They told us we had to ask if we couldn't manage. The lads went logging, Daryl and Mal. They used to bring me logs about every three weeks and we burnt shoes. We burnt anything we'd got, especially old shoes.

Denise: Our fire's done for, we've got to have a new one. We should burn Sunbright; coal, slack, logs, shoes, we had to put anything on to have a fire. My daughter had a cold in the strike and I'd heard at one of the meetings that if you asked the doctor he'd give you a letter and the Coal Board had to let you have some coal. I went, but our doctor wouldn't even give me the letter so that I could try to get it. I was threatened with having the electricity cut off, but we made an agreement to pay so much, about £2 a week. If I missed they sent me a threatening letter. When I went to America for a week I paid four weeks in advance so Malc wouldn't have to worry and they sent a man round to say that we hadn't paid for three weeks. They wanted to cut us off but they didn't manage it. Some weeks I just couldn't pay, when I had to buy bread or something else we needed.

Dot: There were some bad times, and some great times as well. November was about the worst, it was terrible; winter coming hard and men going back. To think you'd served them in the soup kitchen, then you heard they'd gone back. We had a long list of our men, we crossed them off as they went back. You felt terrible, you'd been feeding them one day, the next day they were missing.

Denise: I think it was worse after Christmas. We'd had a good time, all together, the kids had had a good time, a party and presents. People here did well. It hurt us when they went back cos we'd done all we could to keep them out. It was that much worse after all we'd done.

Dorothy Hickling with grandson Adam

Dot: Me, Rita, Linda and Pat, we all got together in the Labour Rooms on Christmas Eve, getting ready for Christmas dinner, doing potatoes, brussels, everything. We cooked three turkeys. We had a marvellous dinner, we had wine. Then they went back to work. I began to think we'd lost when they started going back. We used to ask, "Whose gone back today?" We hoped they'd say none. Women had to leave the group when their men went back. They didn't want to, but it wouldn't have been right, they'd have been talked about, but it was a shame.

Denise: I think the wives did want to keep on coming. I was sorry about that. But I couldn't understand the men going back, they must have known it was going to end. Why didn't they just wait a bit? We had two men sacked at Hucknall for fighting outside the Welfare, not at the pit. I'll never forgive the men who got them sacked, and they got 4 months prison. I'll never speak to the men who caused it. Malc has had to change his shift to work with one of them; he's really bitter.

Dot: She's harder than I am but there's some you'd never speak to again. I won't, as long as I live.

Denise: I'll probably speak to the others and socialise. Men who didn't come out at all, I ask them why they didn't. They don't know, they don't understand, they don't know why they didn't come out. It was for the money.

Dot: And they don't understand they are going to lose their jobs. They're not bothered, not till the time comes. They don't think about tomorrow and they don't think about anybody but themselves.

Denise: I always try to think of two sides to anything, weigh it up, but I could never weigh up their side.

Dot: It was the busiest time of all my life, the strike. The day we went back was dreadful. Suddenly everything stopped. All the new friends you'd made, cut off. All the Yorkshire lads, Kent, Welsh lads, all nationalities, even Americans, women miners; we'd met then all, they'd been here. Everything stopped.

Denise: Yes, you shouldn't think like this but we didn't want the strike to be over. Well, we wanted to win, of course we did, but we didn't want all the other things to stop. It was so quiet after. We'd do it all again you know. I was the only woman in the factory where I worked whose husband was on strike. They used to tell me that principles don't buy food or clothes, they just didn't understand. One night I had maggots, a ton of them, dumped in our garden. The windows were covered and they were coming in under the door. I rang my mother and dad and screamed to them to come and help me. That was the worst thing that happened to me. I don't know who did it but they did it to some other strikers as well. That was terrifying. But I'd go through it all again if I had to, for the strike.

Dot: I went back to the canteen after the strike, back to my job. I'd been the only woman out of six to strike. They felt awkward with me, I know, but I don't feel awkward with them. They don't understand about the union. I do understand. I'm still with the NUM, its my union. I do think there's some men now who didn't strike but who wish they had.

ANNETTE NEEDHAM

Age 30 years, married to Danny aged 32, miner at Rufford Colliery for 8 years; three daughters: Deborah aged 11, Della aged 10, Rebecca aged 9.

We are a mining family. My father died of ill-health at the age of forty three. Danny's grandfather eloped to Notts, galloped up here in a horse and cart, and got a job in the pit. His father also worked at Rufford, that's how Danny got his job after he was made redundant from the foundry. He didn't really want to go down the pit but there wasn't anything else. He was out of work for four years when the little ones were young, it hasn't been easy all the time we've been married.

In January 1984 I hurt my hand badly and Danny had to be off work to look after us. I lost my two jobs, pub cleaning in the morning and work in a fish shop bar in the evenings. It was long hours for little pay but it meant we could run a car, which we couldn't on Danny's wages. We were in debt when the strike started. I tried to ignore it, afraid of facing up to the possibility of it happening to us. I knew what it was like not to have two ha'pennies to rub together and I was terrified of it happening again.

It all came to a head when there was a march in Worksop. Danny had been going in to the pit when there wasn't a picket line and coming home when there was. We were edging up to it for a week or two but I kept thinking that every day was another day's wages, perhaps it would all blow over. He went on the march in Worksop without telling me where he was going and when he came back he was ecstatic. I can remember thinking, "You fool, why don't you show me that amount of emotion?" It's a terrible thing to say, but I felt I wasn't needed, his family was the miners on strike, not us, and his loyalties were with the men on strike, not me, and I was jealous. I couldn't for the life of me see how he would want to put me in that situation again after all the misery we'd been through. Then he gave me an ultimatum:

"Look, you damn well please yourself what you do, but I'm going out on strike and if you don't want me to stay here, then I'll leave." Stubborn as I am, I told him,

"Sod off. If you feel more for them than you do for your family, then go." As he got to the gate he shouted to me,

"You consider you're a fair-minded person, but you haven't heard my side of the story. Why don't you get off your butt and get to a meeting and hear my side." As he opened the car door I shouted back,

"Right, I'll take you on, but if I still don't agree, you'll pack your bags."

We went to a meeting at Sutton, Arthur Scargill was there. There were crowds of people, all for the strike, and I felt odd. I told them when I went in that my husband was out but that I didn't support him. They all looked down their noses at me, but I wasn't going to tell a lie. I heard what Scargill had to say and it really made sense, in fact he said what I'd been thinking for a long time, about rallying round to support each other. I never thought people would do it. The next Friday I went with Danny to the picket line at Rufford. I was the only woman there and then I went with him to the men's meeting in the Drill Hall. I didn't want to go in, with about 120 men, I felt really stupid. Mick Walker was talking and suddenly he pointed at me and said, "That's what we need, we need some women here." All eyes turned on me and I could have crawled down under my chair. Then everyone was poking me and patting me on the back.

I went with Liz and Claire to Blidworth to find out how to do things. I've never been so confused in my life. I didn't know why I was there, really I didn't know anything. They told us we must go out speaking, we must raise money, we must get information; I got more bewildered as they went on. I'd never given that side of it any thought. It was five weeks from the start of the strike and we'd not had any money; we'd lived on the holiday money and it had just about gone. We'd had nothing from Social Security, nobody really knew what we had to do, it was chaotic.

But we did get started at Rainworth. I remember we had no kitchen, no hot water, no kettle, but we cooked soup in an old Burco boiler. I worried about the copper lining of that boiler, I still do. I thought the damn thing might be poisonous. But nobody died. We did the rounds of the local shops, you learned to ask for the children, not for the miners and their families. We didn't get much so we had to start going away from Notts. I went speaking to Hendon and the people there were fantastic.

Whilst I was getting organised, Danny was merrily going out picketing. The kids were starting to get aggro at school. They came running in one day asking, "Dad, are you a scab?" He said he wasn't, he was a striker. They wanted to know what a striker was, so we tried to explain. I talked to them about the union and about supporting men who were losing their jobs. Then, blow me, they were running around shouting, "My dad's a striker." We were trying to keep ourselves to ourselves, it was an awful atmosphere, really dangerous. We thought at that time that we were the only ones round here, then we found three other families and my kids could be friends with theirs.

What got me most was when people I'd spoken to for years and years walked straight past me, or when they pointed at me and said to others, "Her husband's out on strike." It was as if we'd suddenly become like murderers. A friend of Danny's accused him of painting 'Scab' on his house and sending filthy letters to his wife. How could they think we'd do such a thing? They had known us for years and they knew we'd never do a thing like that. They never spoke to us until the strike ended. Then across the road are some other friends, or were. We looked after their house when they went on holiday and they accused Danny of stealing their video. They got the police round to our house. I sat and cried, how could they think Danny was a thief? It never entered our heads to steal anything. But just because we were on strike people made us into criminals. They caught the man who stole the video.

There were a lot of things that got to me. Like the teacher at our Della's school. We had a sit-in at the Drill Hall and the TV cameras took photos. My three were on the film, they had a day off school with me and Danny. The teacher made Della nearly hysterical, cross-examining her about what she had done, telling her she could be taken away from home, put away, that she could be summonsed. She said the strike was stupid. I could have killed her for doing that to my daughter.

When it got near to Christmas there was a lot of pressure, a lot of money offered to men to go back, and letters from McGreggor and the pit manager. I took my girls to the shops and they were outside, looking into a toy shop window. I could see a woman talking to them and I asked them what she had said. They were all excited about their dad going back to work—the woman had told them they could have lots of presents because their dad had only to go back and he would get thousands of pounds. They were asking if they could have BMX bikes and all sorts of things. I told them their dad wasn't going back. It went on for weeks, they kept asking when he was going back. It really dug into me. I hadn't told them a lot about the strike but now I did, I told them everything I knew. They learned the lot, in all its misery, and they changed, they grew up overnight. Then they knew they had a dad to be proud of. I took them to the picket line. They saw the police. They hate them now. It frightens me that they do not respect the police, but how can they, how can they trust them after what they have seen? They saw the police push into our house and arrest Danny at ten o'clock at night. They just told him to get his coat and then they arrested him in front of the girls.

I went on a picket line where I was nearly arrested just because I refused to go home. I've seen men being arrested, being hurt by the police, but the worst I've seen was on the CND march in London when vans of police drove right into the Greenham Common women who were walking and singing, arm in arm. The police knocked hell out of them. I saw it. They tried to push Caroline and me towards the Greenham women, though we were covered in coal stickers. We wouldn't go, and I was screaming at a policeman to leave one woman alone. Another one was swinging Caroline around by her 'T' shirt and I was hanging on to her leg, so I was being swung round as well. I was terrified, the policeman was swearing at us. Those Greenham Common women, they're marvellous people. The police would have beaten us up as well, but we got away and went charging down the road, it was horrible, terrifying. I'll never forget that as long as I live.

I went collecting in London and I got some abuse, but Nottingham was worse. I went to the city stall and a woman spat on my face, then she came back and spat on me again. I could have punched her, but Liz Hollis told me to keep calm, be polite and try to smile. The police were just waiting for us to put one foot wrong and they would have closed us down. I learned my lessons that day. Abuse seemed to get worse every time I went and people would 'accidentally' kick you as they walked by, on purpose.

Things got really bad just before Christmas. We already had debts before the strike and we sold our wedding rings to pay a court fine. We sold the freezer, it was empty anyway, the car went back, the fridge and the electric cleaner broke down. But the worst was the cold. We have no open fire and the stove just heats the radiators, it doesn't give out any heat. It needs five bags of coal a week. We burnt furniture—dining chairs and the display unit; it was old but it would have lasted a bit longer; we burnt shoes; we chopped up our wooden ladder and burnt that. One day we heard of some wood going free on a site being cleared. It was absolutely pouring with rain but two women and four men went down there, two feet of red slimy mud and me in size 9 wellies when I only take size 4. We put on old clothes from the jumble sale stuff, you should have seen us! Even the workmen stopped work to laugh. The wood was twelve feet long by six inches square. We had to toss it over a five foot high hedge, like tossing the caber, running down the slope to get some force up. We slipped and rolled about in the sludge and then the men started throwing mud at us, we joined in and we were like kids again. It was the happiest day of the strike.

The house was freezing cold. We were invited to a friend's house in Newark over Christmas and when we returnd there was green mould growing on the walls. The two oldest girls went to the South of France on a free holiday. That made up for all the other disappointments, they had a marvellous time. At least we were all warm for Christmas. Then I had to go into hospital, for another operation on my hand. I've got a disease of the blood vessels and my hands go blue. I didn't say anything about the disease at the strike centre, I didn't want anything extra. In hospital I was warm and well fed but I couldn't rest for worrying about the children, cold at home, and I went on losing weight. Danny promised me that they would go to his mother's house, but when I came out of hospital I could smell the damp in the house and they had only stayed at her house for a few days. Danny said he had always had a fire but I knew he was lying. The girls all had colds and looked sickly. I lost my temper with Danny. He was trying to protect me by pretending, but I didn't want protecting, I was strong, even though I was ill. And I knew that the house was freezing cold, why did he pretend to me?

The next day he went back to work. It was my fault. He went back because I was ill and I had no warmth. He had talked about going back once in September but I had talked him out of it. See how everything had turned around. He had dragged me into the strike, then I'd done everything for it, he couldn't hold me back. I got on the bus the first week we had a wage again and on the bus was a scab. I couldn't look on Danny in the same way as I looked on that man. Danny wasn't a scab, he wouldn't have gone back, but he thought it would go on for some months more and we were at our wits end. He cried when he came home from the first shift. He cried. It was the loneliest three weeks of my life before all the men went back at the end. We didn't know where we were, no longer a striker, but not really a scab. When it ended so soon after, that made it worse, because Danny said we could have hung on for another three weeks.

MARIE PRICE

Age 50, married to Alan, aged 51, miner at Silverhill colliery. Five sons, one daughter: David, married, miner at Newstead; Kevin, married, miner at Silverhill; Peter, single, miner at Silverhill;

Mark, single, miner at Silverhill; Colin, married, plumber; Tracy aged 13.

Alan has worked at Silverhill for 30 years. My father and grandfather were miners, my brother is a retired miner. Alan's father was a joiner for the Coal Board, his grandfather was killed when a cage fell on him. My own dad used to talk about the Spencer Union, the boss's union, and he was very bitter about what it did in Notts.

After the first time the Yorkshiremen came down here to ask our men to support them in the strike, I went shopping in Sutton with Alan and met some more Silverhill miners. They said no Yorkshireman would stop them going to work and we argued about it. Then, the next day, Alan told me that he'd heard of men sitting in their cars in the car park at the pit all day, frightened to go home and tell their wives they had come out on strike.

Alan agreed with the strike but went along with the Notts men who wanted a ballot. I didn't agree with him and things were very bad here, we nearly had a divorce. It was making him ill, going against what he believed in, we couldn't be in the house together, he walked in and I walked out. Then one day he said we'd got to have it out and asked me what I'd do in his shoes. I told him I'd be on strike. That settled it and he said we had to talk to our Tracy, she would have to understand there'd be no money for a month or more. Well, we'd never had much money, so it couldn't be that bad, and Tracy agreed. Alan came out just after Easter and Peter, Mark and David came out as well.

Very soon after that, Arthur Scargill came to Sutton to speak and I really did want to go and hear him. But I have athritic spondylitis and Alan thought I might get hurt in the crush. There was a big crowd and I might have got hurt, but I still wish I had gone. Alan came back from that meeting with Gwen Macleod's phone number. She wanted to get some women together to start a group. There were about eight women at first and I was with Janet Gibson in a little kitchen in the Hillocks Complex from about 7.00 am till 12.00 noon. Janet came in after she had taken her little girl to school. We did mainly breakfasts for the pickets. Then we got the Community Centre at Westbourne School and some more women came. I was there from 7.00 am till teatime, buying and cooking all the food, then going home and doing more cooking ready for the next day at the centre. After a while we got a rota and three shifts and that was better, but I was still buying all the food. I collapsed one day in Sutton and had three weeks in hospital, but after a week home I went back to the kitchen.

I thought that would be my work for the strike but someone asked me if I would go out speaking. There weren't many women of my age, most had little children so couldn't go away from home. I had never gone out anywhere except shopping; my married sons always knew where to find me, always at home. But now I had no excuse and I said I'd go to Brighton with another woman and two men, just for company. I ended up speaking the first day at Sussex University, to an auditorium full. Then I did one after another, meetings till 8.00 at night and I never flinched. I never minded, because I knew what I said was all true and I believed in it. At one Trade Union school a man from Matlock stood up when I had finished and said I had converted him, I'll never forget that. I went on doing meetings, I would go away for days, all over the country, to London, to the Westminster group of teachers, to Weston-Super-Mare, to Lincoln, to Scarborough and to Richmond. I went to Germany, a different town every night and others in between. I hadn't been in a plane before and never abroad. I hardly saw my daughter, I was never in. But she didn't complain and I took her with me for a week to Westminster, it was the only way we could be together. I think it did her good, she learned a lot, I wish I had known what she knows now when I was her age.

I think my life had been like one long lie. I am 50 years old now and I have been living in cloud cuckoo land. I'm ashamed of my life because I taught my children lies. I told them that the law was there to look after them, so they should not break laws. I told them we had freedom in this country. Well, we do, until we do something they don't like. They told us in Notts that if we were strikers we had no rights, the police told us that and they proved it. They were everywhere on this estate, on every corner and in every cul-de-sac. We took the dogs for a walk across the fields at the back here and they were running into the hedge bottoms and sniffing out the police; they were popping up like mushrooms, black ones. I think my upbringing was wrong. Because I was a girl, teachers told me I couldn't do things boys did, it wasn't ladylike to be a tomboy. I push Tracy now to be equal but she doesn't need pushing. I know women can do all the things men can do and I'm sure we do some things better.

This was a male chauvinist house and it was my fault really because I brought up my sons like that, women still do. I used to stir the sugar in Alan's tea. Now he says I've left him behind. Once at a meeting a teacher asked him if he was proud of me. He said, "Well, I think I am but I must have lived with a stranger for fifty years and I'm not sure which woman I like best, old or new." One night I rushed home to change my clothes before I went out again and he said, "Hey, this is my bloody strike, you know." I said, "Well, is it?" He says that when he married me I was behind him, then he blinked and I was in front. But he knows we are closer now than we've ever been.

The strike was the first time I'd ever been on a picket line. I only went twice, I didn't have time. I stood where I wanted to stand, on the pavement. Straight away a policeman told me I couldn't stand there and I said I could. Alan was on the line with the men and he kept shouting to me to stay where I was, it was my right. The policeman kept shouting to me to move. The two of them going on at me, telling me what to do. I was sick of them, so I told both of them to shut up. I would do what I wanted to do, I didn't need them to tell me.

My son David only stayed out for thirteen weeks. His wife was expecting a baby and she was getting herself into a state, so he went back. Kevin never came out. He said that while union men went on working, drawing their money, he would do the same. I argued with him and told him what I thought of him, but we did not fall out. I gave the strike a lot but I wasn't going to give my sons. He just came to see me now and then. Peter was living at home; as a single man he got nothing except his picketing money, but he stayed out all the time. Mark was just seventeen, living at home, and he got nothing at all as the men said he was too young to be picketing. He went to live with the family of his girlfriend in Shirebrook. Then one day Alan came into the centre, his face green; he looked awful. He told me Mark had just passed him on the picket line going in to work—and I had given him my last two pounds only the day before. He stayed away till after the strike was over, but he is back home now. Colin agreed with the strike but he thought it was degrading and disgusting for me to work to feed the men. Now he won't let me see my lovely little grand-daughter. I went to take her a birthday present the other day and I had to bring it back.

We lived on £20 a week, for five adults, but we never went hungry because we had so much help. We had help from people I never thought much of before. I had never even met a black person, now we stay with them in London and they are my friends. I met some marvellous women. I never thought I could beg but I did and I still do, for the sacked and imprisoned miners. I would do it on my knees if I had to. How can we just leave them to go on suffering?

TRACY PRICE

Age 13 years, daughter of Marie and Alan, miner at Silverhill Colliery for the past 30 years, four brothers miners.

My brothers and my dad explained the strike and I learned for the first time about being in a union. My mum had never done anything in her life except look after us, her life revolved around her family. My dad always believed in unions and was interested in politics, but he never talked about these things at home. When the strike started it changed everything. Our house was full of the strike, it took over everything. Especially it took over my mum. I hardly ever saw her because I was at school and when I came home I had to get my own tea most of the time because she was at the support group kitchen or out speaking. I didn't mind and I never grumbled because I knew she had to do it.

I went to the centre at Whitsuntide when I was on school holiday. At first it was strange and I didn't know anybody, but then I got to know lots of people and some my own age from my own school whom I never knew before. Sometimes I talked to the pickets and they talked to me, or else I listened to what they were saying. It was quite exciting when they came in, but in between things were quiet and dull. In the summer holidays I enjoyed it more because by then I knew everyone, lots had come to our house. I helped mum in the kitchen, making tea, or sometimes, when they were very busy, cooking the whole time. I also looked after the young children whilst their mums were cooking or cleaning up.

At my school some boys argued with each other about their dads, being out or scabbing. I argued with them, but it was never really nasty. I tried not to talk about the strike to my friends, but it sort of made things cool between us sometimes. At Christmas my friends were saying what they were going to have for presents, and I knew I was not going to have very much. They were telling me they were going to get leather box jackets and computers. I told them that I had something they could never have, a dad to be proud of. When I told my dad this, tears rolled down his cheeks. Then he said to me, "I think I shall go back to work tomorrow." and I told him that you can't get wheelchairs down the pit. He said he hadn't got a wheelchair. I told him he would have one if he tried to scab. I knew really that he was only joking and he knew that I was sure he would never scab. As it happened I had the best Christmas I have ever had in my life, a sackful of lovely presents that came from all over the country. There was a party at the centre, we were all together and it was really good.

My mother went to London to speak at meetings for a week and I went with her, so that we could have a week together. I never thought she could do what she did, but I think I might be able to do it as well now. The group was Westminster Teachers' Support Group and the people talked specially to me to find out what my opinions were. I also went to stay with people on strike at Betteshanger in Kent, with about 50 others from this area. They wanted us to tell them about how it was in Notts. I had a holiday with 16 young people from this area in Rotterdam. I had never been abroad before and I thought it was great. They were all in favour of the strike and they were working hard for us.

It was a marvellous year for me, exciting. I met people from all round the country and I do want to go abroad again. I know that I did not have lots of things I was used to having. I had second-hand clothes, but they were nice. My mother used to call me 'Gimme, can I have' before, but in the strike I knew it was no use asking and anyway I didn't want them to feel guilty about saying no.

The worst time of all, and of my life, was when I went to the rally for the miners in London. It was 24th February and I shall never forget it, I was so frightened. We all got separated from each other and the police had horses and were riding into people and knocking them down. They kept pushing us and wouldn't let us walk to the coaches to go home. I cried and other children cried as well, then their mothers cried, it was horrible.

I used to watch TV and I didn't think they told lies, but now I know they do, we have proved it. I know that the government wants to break all the unions, they broke the steel industry. I have learned that and I have learned how marvellous it is to have good friends.

MAUREEN (MO) WHETTON

Age 44 years, married to Paul, aged 45 years, miner at Bevercotes Colliery for 30 years; two sons, two daughters: Daryl aged 19, Karl aged 16, Clare aged 14, Michele aged 9.

Paul has always been active for the Union. In the year of the strike he was Branch Secretary at Bevercotes. Before that he had been on the union committee for years. He was secretary of the Nottinghamshire Miners Forum all through the strike. In 1983 he went with a delegation from Notts to Moscow and Leningrad to visit coalfields and learn about unions in Russia. In 1970-71 he went to the London School of Economics to do a diploma in economics, full-time. He was often out on union business and at day and week-end schools.

I have always liked to work, for the bit of extra money, to meet people and to have my independence. When the strike started I had a job working 13½ hours in Tuxford at the fish shop. Before that I used to work behind the bar. My granddad and all my uncles were miners and my dad is a retired miner. It was the same with all Paul's family so we are a real mining family. My son Karl starts his apprenticeship at Bevercotes in September this year.

Paul was at the Welfare when I got the news that Cortonwood was out on strike. Henry Richardson phoned to say that the Yorkshire lads were coming down to Notts and would I let Paul know. He went straight down to Bevercotes to tell the lads and he never went back to work any more until it was over. He didn't have to ask me at all because he knew I agreed with it. I didn't think that it would last long, though, a few days or a couple of weeks. When we saw that it was going to be a while we got DHSS forms and claimed what we could. I cut my hours at the fish shop down to four but then I decided that I liked my job and I wanted it back when Paul went to work again, so I started to do thirteen and a half again. We managed fine, though we went without a bit. If it was something desperate like children's shoes, it would be Shelley's turn first, then we'd save up for what Clare needed most, then Karl, then Daryl. He was unemployed till he got the dole in September, so I was keeping four children. They all understood. I don't think it was hard, in fact it was not hard at all; we were happier than we'd ever been, believe it or not. We had no problems. Even when you are working you sometimes have to say no and there's a pulled face. The children accepted it. I was happy, it was a good year. If the miners had got what they wanted it would have been better still.

I met a lot more people, different people, and I went out more. I went to meetings when I could get there, but being in Tuxford, in a village, it was awkward. I went to a day-school with Labour Party women and striking miners' wives and that was good, we all had a lot in common. I've been in the Labour Party for about eight years now. It is always politics in our house, but during the year of the strike it was even more so. The best time for meetings was at the Labour Party Conference. I was on my own most of the time because Paul was the delegate for Newark. I watched it in the video room, on my own, but I didn't mind. The evenings were best, the meetings and seeing all the people. They asked me for my opinions. I wouldn't have believed it was so marvellous, I can't tell anyone how much I enjoyed it. I got more confidence somehow.

Paul used to go out every morning at about half past seven. He went over to Ollerton with Gordon Darlington, to the strike centre. Occasionally he came back at half past four, but mostly he would ring up and say that he had to go to another meeting, perhaps to Sheffield. I saw him when I saw him. He was away most weekends, speaking, at Basingstoke, Norwich, all over the country. Sometimes it did get me down, Paul being away so much of the time, but I'd soon get back up again because I realised he had to do it, and I wanted him to do it. I had the children, so I wasn't on my own. If it had been just Paul and me I think it would have been harder. When he had to stay away overnight a few times together he would ring up and say, "I'm sorry", and I would say, "Don't worry about it, I don't mind." And I really didn't mind, I wouldn't have liked it if he had been just sitting at home waiting to see what was going to happen. He knew very well that it was alright by me. It would have been exactly the same if I'd have wanted to go away to Durham or South Wales like some of the women did. I'd have gone, and I wouldn't have asked, just told him I was going. We've been like that all the time we've been married. I'm no housewife, I do what I have to but no more, and Paul knows, but he never tells me what I should do. He knows it wouldn't be any good if he did.

I went with him on nearly all the demos and marches and we took the children. I did enjoy them, it made me feel good when we all got together, we were in the same boat and we were all cheerful. I desperately wanted to go picketing, but again it was transport. I used to get phone calls, discreet of course, telling us where to be, but there was never anyone to take us. There were only four married men and five single lads on strike from Tuxford and only Gordon had a car. He needed that for pickets. We had to get to Sutton for five o'clock in the morning and I never did get there. That upset me, I really wanted to go and see for myself, and I wanted to shout. Perhaps it's as well I didn't go, I should probably have got arrested.

It was hard for me to get to our kitchen in Retford but I managed to go to Ollerton sometimes and work in the kitchen there. The women at the central group would phone me about meetings and food collection and I would let the Retford women know. They were nice, the women at Retford. At first I was their delegate to the Nottinghamshire Women's Support Group meetings, and I enjoyed that, getting to know the other women and what was going on. But we were out in a village, strangers, and after a while I think they wanted their own Retford delegate. I didn't mind, but I missed the meetings.

At Christmas we got extra food parcels and presents for the children and Shelley went to the party at Retford. Our own Christmas was one of the best we've ever had. Daryl made home-made wine and beer and Paul's sister and mother came to stay, just as always. Of course, we were very careful who we invited; this year we really picked our friends. We had a fantastic time. Paul's sister was really good to us in the strike, and mum and dad helped us as much as they could. Dad is on a pension so hasn't got a lot, but there was always something for me or the kiddies when we went.

When Paul came home we wanted a rest from the politics of the strike. We would comment on the TV but we didn't talk unless I wanted to know something special. Then he would always explain, but he has always been like that. He wanted a bit of peace and quiet when he came home. He was always tired, but he enjoyed what he was doing, his heart was in it. He was such a happy man in the strike, I don't think he's ever been happier. When the end came he knew he had to go back to work but he will keep on fighting, he is still going to speak at meetings.

When the strike ended I can't tell you how I felt. I always thought we'd win and so did Paul. It was when the Yorkshire men started to go back; I thought, 'surely not, they won't be the strike breakers.' But as time went on, and they went back, never mind how much I was hoping and praying to win, I knew we were going to lose. Paul didn't say a thing.

I changed in the strike. I appreciate things a lot more than I did before. It did me good. I'm not cut out to be a housewife and I could never make my money spin out, but I'm a lot better now. We had a good year.

Paul Whetton was sacked in February 1986 for pinning up notices of NUM branch meetings at Bevercotes Colliery.

YVONNE WOODHEAD

Age 35, married to John, miner at Blidworth Colliery for seven years; three sons: Colin aged 17, John aged 10, Michael aged 9.

I have never worked for the NCB but the strike seemed to be as much my strike as John's, almost more mine than his really, because I did more varied things than he did and I think I did a lot more too.

I was born right opposite the pit at Blidworth, my grandfather and father were miners there, John's father worked at Welbeck. His brother-in-law is a miner at Annesley and my two brothers are miners, both at Blidworth, so you see how strong my connections are with mining. But of all of them, John was the only one in favour of the strike and the only one in the family who came out. In March when nobody in Notts was really sure what was happening, John went to work one day, stayed home the next, went to work, stayed home, refusing to cross a picket line. A lot of the Nottinghamshire miners were doing the same thing, just going in to work when the pickets were not there. John and I did not really discuss the strike issues but I told him finally that he had to make up his mind, in or out, this was neither one thing nor the other. I'll never forget when he said he was going in, picket or no picket. I drove him to the pit, got out with him, and together we looked across the road to the picket line; one or two were Blidworth men, already out, but most were strangers. I asked him again if he was going in and then I watched him walk across the road. I wanted to scream out to the pickets, "Don't let him pass, stop him!" Then I saw what he did. He just opened his plastic bag of sandwiches, gave them out to the men on the line, and stood with them. That was it. He was on strike. It was great. I was so pleased that I wanted to shout to everyone and tell them. And never once in all the time did he mention going back.

It was to be my birthday at the end of March and John said that there would be no birthday presents this year. I didn't care about that, in fact I was quite proud about it. I told him that the strike would not last more than about six weeks, anyway; that miners have quick tempers that flare up and die down; I told him not to worry. We had a bit of savings and we could manage for six weeks. I began straight away to cut out cakes and biscuits and boiled ham, to make the money go further. After four weeks our savings had gone and nothing was yet organised in the village for us. I had three rings from my first marriage, engagement, wedding and eternity. I sold them all and the money kept us going until the supplementary benefit came through, £38.50 for five of us. We had been used to an income of about £90; the overtime ban had been on for about six months, but anyway, miners' wages are nothing like as high as some people think they are. I did wonder how ever we were going to manage for a strike that looked all set to go on longer than we thought in the beginning.

Then we got our support group going in Blidworth and the centre opened at the Village Hall, the 'Glasshouse' we call it. They had to sit in to get it and it became like our home for the rest of the strike. Because John came out early I was involved right from the beginning and I began to learn more and more every day. One of the first things that we had to do was to start raising funds for food, and women began to go out speaking. John didn't want me to go, he was really opposed to it. So I didn't let him know when I was going. I told him when I had been and then it was too late, what could he say when I had done it anyway? I went out locally, when he was safely away in the pub with some of the other strikers. They would discuss things for hours so I knew I was safe and would get back before he did.

He never thought of going out fund-raising himself but once when he was feeling a bit low I persuaded him to go with a group to Cambridge. We were twinned with a group there and I knew he would enjoy it. So I pushed him out and it really did him a lot of good; after that he went again and again. We used to sit together and watch every news broadcast and every programme about the strike. Before the strike, we watched, but never really made any comments on news items, just grumbled a bit about unemployment reports and other things we didn't agree with. But we never before looked properly at issues and talked together about them. I learned so much, especially from the other women at the kitchen, and I learned from all the visitors who came to the Glasshouse, from all over Britain and from abroad. They would ask us questions so we needed to give them the proper factual answers, it kept us on our toes.

I picketed the pit in my own village where our own miners were gradually seeping back. I learned about police tactics early on in the strike. They were lovely to our kids at first, playing cricket on the grass when there wasn't a picket on and giving them sweets and crisps. Then we found out that they were asking the children where their fathers were going to picket the next day. After that we began to code everything to prevent them getting prior notice and putting roadblocks on. Before we got the Glasshouse for our kitchen we used a house for our pickets to meet and have breakfast. As soon as they found out, the police sealed off both ends of the street and only people who actually lived there could go through. My friend, Betty Savage, lived there and I couldn't even go to see her. I have always gone to see my father every day, he lives just about ten minutes walk away. One day I was stopped four times, asked where I lived, who I was and where I was going. If they did not believe our stories we were turned back. Blue uniforms spread all over our village—our village, not theirs! I was furious.

But the worst time of all with the police was when we were on an all-women's picket and Mary Dickey from Blidworth was arrested because she refused to trample over a woman who had been pushed down. We intended our picket to be a peaceful one, we always did, but the police were in line after line, behind each other, and all round us, trying to heave us down the road to stand in another place. We didn't want to be moved, we were doing nothing to hurt anyone, just singing and shouting, all in good humour. The police pushed and shoved and elbowed us, and it didn't matter who fell down, they just kept pushing. I am disabled and I lost my walking stick. I wouldn't go where they wanted us to go so the Inspector spun me round and threatened me with arrest. I still wouldn't move and in the end, when they saw I would not give in, they let me stay and five more women were allowed to join me. So we won that one in the end, though it was dreadful for Mary, locked up in the police cell. I have never had anything to do with the police before that day and I never want to have anything to do with them again.

The worst time at home was in the winter. We had no coal and no heat. There was no hot water except what we boiled in the kettle. The whole house felt damp and chilled through. At night after we had been to the Glasshouse the boys would sit together in their sleeping bags whilst they played cards. We had a hot water bottle that we shared. We passed it round in turn and hugged it for a bit of warmth and comfort, even visitors who came to our house shared the bottle. One night as I filled it, it burst in the sink. I couldn't believe it, I just stood and watched the precious hot water draining down the plughole. The worst part was telling the kids and we all cried. I have kept that bottle as a souvenir; if I could have it mended I would. Our car had been smashed up and was off the road, supplementary benefit had been reduced to £25.44 because my son had left school, but nothing seemed so bad as our bottle bursting, losing that bit of warmth and comfort.

As the strike went on and on and as Thatcher deliberately made things worse for us, our supporters gave us more and more to make up for it. The Tories could never break us because they didn't understand that poverty kept us close and when we were close we were strong. One person's problems were everybody's, we wouldn't let anyone be hurt if we could help it. All my friends now were part of the strike, the others have gone. They ignored me, wouldn't look at me in the street. I don't see them any more. It was specially hard in my family. My dad has always been my best friend and my mother died as the strike started. He was against the strike so in his house I never mentioned it, but when I saw my two brothers who were scabs, I used to shout out, "Scab, scab", to them and I enjoyed it. All the members of my family thought John was stupid to be on strike, losing so much money. None of them helped us, but we managed and we don't owe them anything.

The most interesting thing that I did in the strike was to work in the Nottinghamshire Women's Support Group office that NUPE lent to us for all our organisation. Betty Savage was working on holidays for Blidworth group and she asked me to go into the office with her to make a change in arrangements for my two youngest sons. The women there were so busy that they asked me if I could spare some time to help out. I was worried at the prospect, I had never done anything like office work and never seen a typewriter, even a child's, before. But with practice I became really efficient; I kept files, typed reports and letters and had contact with all the support groups in the county. That way I knew what was happening to all the women, I saw the whole picture rather than just my own bit in Blidworth. It was so interesting and so rewarding for me that I would have gone there seven days a week if I could. The work kept me so busy that I didn't have time to worry about my own problems. The arrangements for holidays for the children was my main work. It was quite a worry because it seemed that if everything went smoothly beforehand there were problems with the actual holiday, and vice versa. I worried too about sending children away to the care of people we did not know. But everybody survived and had some wonderful holidays. I was lucky and went to Marseilles with a group of children. I had never flown before and neither had I been out of England. Never before had my boys been away from me and now they went to a different part of France. All the French people were so kind and could not do enough to give us a good time.

I feel that I gained so much from the strike. People say that we suffered. It depends what you mean by suffering. We had to do without lots of things we liked and were used to. The worst thing was being cold and not having hot water. But I think we gained so much in exchange and I would not have missed it all for anything. I have changed so much. I used to think that politics was for politicians, like any other job, it was for them to make decisions and do their work. I was not involved except just to vote at election time. I never questioned what I heard or saw, I thought reporters wrote facts and I thought pictures on TV were just as they were taken. But in the strike I saw for myself how facts can be used, and pictures too, twisted round and round till they come out differently, as lies, the opposite way round. I didn't know that we had a Labour Party branch in my own village before the strike, but now I have joined. I want to go on learning, especially about politics. And I would like a full-time job that interested me and seemed to be worth doing, like the things I did in the strike.

THE PIT SUPPORT GROUPS

The following accounts of the pit support groups were given to me during conversations with groups of women over a period of about three months immediately following the end of the strike. Some of the groups with whom I spoke were large and some small. Some women chose not to join in this activity at all. I understood that many were desperately tired by this time, in some cases near to exhaustion, both mentally and physically. There was mounting anxiety over presures to deal with debts and a wish to reestablish a familiar pattern to their lives and to provide some stability for their children. Yet, despite all this, many women willingly gave me their time and searched their memories for details and emotions associated with events already beginning to blur into a general mass of experience. They all regretted that they had not had the time to keep diaries during the dispute.

If names of women or supporters, or descriptions of events, have been omitted, I know that the women would want to explain that this is inadvertent, part of the problem of remembering details from so much that happened to them during the year of the strike. Inevitably accounts will relate to the perception of the women who gave them and will not match exactly the way in which others present might have described them. I believe, however, that any mistakes or inconsistencies will be small and will not change the essentials in any significant way.

Most of the letters from supporters reproduced in the chapter were received by the central group, but they are typical of many thousands which poured into all parts of Nottinghamshire during the strike. I have chosen to include mainly personal ones which express the warmth, humour and understanding felt by supporters from all parts of Britain.

ANNESLEY AND NEWSTEAD WOMEN'S SUPPORT GROUP

On May 15th some of us went to an all-women's rally in Barnsley. On the coach we got talking to Gwen, Sharon and others from the Sutton-in-Ashfield group and they told us what they were doing and how we should set about starting up our support group. We couldn't wait to get home to get started. On the Tuesday we arranged a meeting at the Annesley Welfare and got word round for women to come. There were about ten of us that morning. We decided that the first job was to get somewhere to use as a kitchen. Christine Winnals said she would ask some of the men on the Welfare Committee if we could have the Welfare. At first they agreed, then they must have thought again cos they took it off us after just one morning.

Next we tried for the Welfare at Newstead. We were luckier there. We got the place for two hours from six in the morning, just for breakfast. I suppose that was something, but compared with what we needed it wasn't anywhere near enough. The biggest number of strikers who came for breakfast was about 80-100. They lived in this area, though they worked at Calverton, Gedling, Blidworth, Moor Green and Church Warsop, as well as at our two pits. We kept really detailed accounts of how we spent our money.

The books are very clear. They show that the first week's income was £30 and the expenditure was £30.20. The first money coming to the group was from the official union hardship fund and the central women's group. Chris Winnals and Jane Simpson went round all the local shops asking for donations of food.

We got the magnificent sum of forty pence and half a bag of potatoes, and we had to go round to the back door for that so that no-one would see the shopkeeper giving anything to the strikers. At one shop in Kirkby-in-Ashfield one of the women asked if we could buy day old bread on a regular order. The manager told us he gave it to feed the pigs. That's what we were up against in Notts.

Some shopkeepers, however, showed their support by making their prices as low as possible, "They knew they'd want our trade again when the strike ended."

The first money raised specifically for the group was collected by Brian Walker, a striker and Branch Secretary from Newstead. It came from various local Labour Party branches as well as from some individuals. Vera Darlington from Hemel Hempstead was a constant supporter, "She sent us money and food right from the beginning and never let us down. We owe her a lot, we do." At first the men went out collecting regularly and gave money to the women, but later the women decided to collect themselves.

> We went to Birmingham and we divided into two small groups on opposite sides of the road. It looked to my group as if those on the other side were getting a lot more than we were. We couldn't understand why. Then it suddenly dawned, they were wearing our Notts women's 'T' shirt with the slogan 'Notts Women Strike Back'. Folks obviously thought this group was against the strike! We didn't know what to do for a minute, then the others covered themselves up, we didn't want anybody thinking any of us were scabs' wives.

Food parcels could not be packed in the Welfare. At first it was all done at Evelyn Stanley's House, then later at Chris Simpson's.

> We tried to put ten items in every parcel, even if they weren't always worth the same. People got to look forward to ten things. When there was a family with a lot of kids we tried to give extras and if they had a new baby we made up a baby basket, with everything they needed.
>
> ★
>
> The group always thought a lot about the children, the strike didn't seem fair on them so they had special treats. We gave them a 'goody' bag every month. There was an orange, an apple, a bar of chocolate, crisps and sweets. It was good for the kids and good for their parents, they felt that they weren't missing out quite so much. I remember when one of our young strikers was left alone, his parents went off on holiday leaving him with nothing. That was a terrible thing to do so we gave him a 'goody' bag and it cheered him up a bit.

Days of the week will always have a special significance from now on: Tuesday—sorting food left over from last week's parcels and making new lists, Wednesday—shopping, Thursday—packing, Friday—distribution.

Packing food parcels week after week was an onerous task, but it was very important to keep up supplies. To cut down on the boredom the women tried to make a game out of it, seeing who finished first.

It wasn't very funny having sacks and sacks of potatoes in your front room either, especially when they started to smell. You couldn't move for the damn things.

★

We were getting more support as time went on and we made lots of links with trade unions. Newstead and Annesley men went speaking to St. Alban's and Croydon and we got fantastic support from people there.

On September 13th, 1984, a letter came from Moss Evans, the General Secretary of the Transport and General Workers' Union, expressing support. This was followed by a donation of about £2,000 worth of baby food.

We'd never seen so much baby food all together. We couldn't cope with it all, tins and tins of Ovaltine, rusks, nappies, those little jars of baby food and such like. We took what we needed and gave the rest to the central group to distribute. They were glad of it.

Christine went to speak to a group of lecturers at Nottingham Polytechnic:

I was so nervous I couldn't remember one word of what I said. But I took a typical food parcel to show them what we had to live on and that was a good idea. You could tell they were shocked at how much we had for a week and they said they would help us as much as they could.

Strange food came to all the support groups at one time or another, from France, Germany, Bulgaria and the Soviet Union. Some people really liked the strange 'concoctions', others wouldn't even try. They had fun trying to guess the contents and sometimes the only thing to do was to tell by tasting. One day tons of tampax arrived from Croydon. Someone had obviously realised that women needed such essentials and had bought in bulk. The men were a little embarrassed at having to collect them and take them home for their wives.

> We managed to have a lot of fun, even when times were really bad and we were all worried to death. I don't know how we made jokes when we were up to our eyes in second-hand clothes, but we did. We used to go through the pockets hoping somebody had left some money and once we found two pounds. That was a fortune to us in the strike. And there were some lovely clothes.
>
> We had a stall in Hucknall market and we sold everything we could lay our hands on, we haven't got many bits and pieces left, everything had to go.
>
> I think the best times were in the early morning, down at the kitchen, having a good natter—as long as the stove didn't blow up. It was overworked, that stove, and it kept needing new fuses. We used to cook sausages from big catering tins, they tasted really good at that time in the morning.
>
> Christmas was something special. SOGAT brought five hundred turkeys and five hundred toys and books. We didn't need all that so we sent the surplus to the central depot at High Pavement in Sutton. Dave Roe and Ken Williams from Central TV sent us £400 for food to be fetched from the Cash and Carry. We did special parcels for Christmas at Hilary Whitehead's house:- lemon curd, jam, mincemeat, jelly, fruit, nescafe, corned beef, peas, sugar, pickles, 3 lbs flour for baking, stuffing, all in a special Christmas bag. We had a turkey and one big and three little presents for every child and a pound bar of chocolate. Just think what that meant to everybody.
>
> ★
>
> Well, we knew it was going to be important so we started putting

Brenda Prince/Format

Doreen Brierly, Striking Canteen Worker, Newstead

bits by and planning, right back in September. We planned a party with Father Christmas and a big lucky box for the kids. When it came it was the best time you could ever imagine, we'll never forget it as long as we live.

There are some bad times to remember too, as when a woman had nails put in the tyre of her car and it blew out when she was driving children to the nursery. The same woman was maliciously reported to the NSPCC and had her clothes line cut so that all her clothes fell onto the muddy ground. One morning police tried to stop women on their way to the Welfare kitchen in the early morning, even though they were so obviously carrying bread and other food for breakfast. They get angry when they remember all the police activity or when they think of the time the local pub landlord and landlady accused them of causing trouble and intimidating other customers:

> We annoyed the scabs because we were in such good shape and we always wore our badges. That was all the trouble we were causing. We blacked the pub for a bit, but we're back there now. We all sit together, all the people from the strike, and we have the best time of everyone. They can't stand it cos we're so happy.

The strike made us strong. We sing in the pub every Saturday night, in defiance, "We are women, we are strong" and we stand up for ourselves like we never did before. We've got our own points of view now and we can defend them.

And we'll never forget those who helped us.

Women Involved in the Group:

Committee:	Chris Simpson and Janet Allen from Annesley, Evelyn Stanley and Hilary Whitehead from Newstead
Annesley Women:	Christine Winnals, Aurea Chisholm, Rose Hurt, Debbie Stevens and her mum, Mrs. D. Roe, Tricia Concannon, Jane Simpson.
Newstead Women:	Jeanette Hackett, Doreen Brierly, Cassie Stanley, Alice Walker, Marilyn Lunn, Olive Kiddy, Elaine Gascoigne.

Supporters:
SOGAT, NALGO members from Sheffield, People in St. Alban's, Croydon, Luton, Forest Fields, Central TV and the BBC, Vera Darlington from Hemel Hempstead, Pretty Polly factory workers, staff at Berry Hill, Lecturers at Nottingham Polytechnic.

ASHFIELD WOMEN'S SUPPORT GROUP

We started in March. We were one of the first groups, with about 120 miners on strike living round about and normally working at Silverhill, Sutton, Bentinck and Babbington. These pits are all about three miles or so from the town centre. The number of strikers dropped to about 75 at the end. At first there weren't many women, but as time went on and other people heard about us, more came along, so it was better, we could have a rota. We used to do breakfast about 7.00 am,then we'd be open from 9.00 till 5.00. Later we cut down and finished at three o'clock.

At first we used Hillocks Community Hall in Sutton, but it wasn't adequate for what we wanted, so we asked the District Council if we could use the Westbourne Centre, which used to be a school, and they let us have it free of charge.

Like most centres it was more than a place to come and eat, it became a social centre, like a temporary home.

We loved coming down here, it was a place where you felt you were all together, sharing what you'd got, getting practical help if you needed it, and advice on welfare. We all got on together, we had joint meetings with the men and we worked together well. They were good, they didn't resent our independence, they thought we were great. They shared the work here and they helped us at home, even if some weren't too keen at first. They had to get used to doing it when we were down here or away speaking.

Without this sharing of domestic and child-care responsibility the women could not have done so much. They were a very active group, full of vitality, involved in all aspects of the strike. Whenever there was work to be done or solidarity to be demonstrated, the women from Ashfield were there. They were always at central group meetings in Nottingham or Mansfield, regularly on picket lines, speaking at meetings throughout the country and abroad.

One of the group members, Gwen Macleod, was already knowledgeable about trade union matters and politically aware, and she gave impetus to the others. At the time of writing she is the chairperson of the Nottinghamshire Women Against Pit Closures Group, formed out of the central group.

The Ashfield group also looked after the central 'pick-up' point on High Pavement.

> This house was empty and the District Council lent it to us to store everything till it could be distributed. It was always full of clothes, surplus things from other groups and at Christmas it was full of toys and food. It was extra work for us but it was great to have it.

The Ashfield women raised an enormous amount of money by going out speaking and sent it all to the central group. At the very first fund-raising march in Worksop they collected £24 of which they immediately sent £8 to central funds. Sharon explains:

> We had our own money, for our kitchen, from regular donations. Gwen wrote a letter right at the start to all the local trade union branches. She got in quick and that helped our funds, but we needed much more than that. So we sent letters to branches further away.

The joint activities within the group were really enjoyable, helped morale and made all the hard work seem lighter, so that women will always remember the strike with very positive feelings. They will never forget how they cared for each other, met new challenges, and had such success in their ventures. Neither will they forget the opportunities they had to share their experiences and solidarity with strikers and their families from the other coalfields. Writing in *Here We Go*, the magazine of the Nottinghamshire Women's Support Group, Sharon Reed describes the joy of travelling to the Scottish Miners' Gala in Edinburgh with three coachloads of families from Nottinghamshire and Derbyshire (Bentinck, Sutton, Silverhill, Pye Hill, Moor Green,

Babbington, Derbyshire Workshops, Mansfield and Clipstone.)

> The Scotland trip came about after two members of the Ashfield group had had some pickets from Scotland stopping with them. The Cardowan Branch invited us up for the Gala. They thought it would be a good idea to show Scottish people that there were Nottinghamshire miners on strike. The reception we got was overwhelming and brought us all to tears. We stayed with the Cardowan miners, and went with them to Edinburgh for the Scottish Miners' Gala. We took the Notts area banner for the march. When the Scottish people saw it they went crazy, applauding, hugging and kissing us. The emotion was so great that we all said that we would remember the trip until the day we died. We didn't have to pay for anything, there was as much food as we wanted, in fact there was just about anything that anyone could have wanted.

The Ashfield women presented a silver plate to the Scottish miners to thank them for their hospitality; on it was engraved:

From Notts striking miners
in friendship with Cardowan Colliery

Ashfield Women's Support Group

The women continue to meet as a group as they recognise that there is still work to be done for the sacked and imprisoned miners. They are also very active in the fight against the break-away organisation in Nottinghamshire, the Union of Democratic Mineworkers.

> We also want to pay back all those who helped us by being ready to raise funds for any other group of workers who are attacked like we were.

Women Involved in the Group:
Committee:
Chairperson: Joan Minett
Secretary: Gwen McLeod
Treasurer: Sharon Reed

Members: Lou Dove, Janet Reed, Marie Price, Ann Cassatari
Other Women: Julie Starr, Sheila Brooks, Sharon and Brenda Fletcher, June Chapman, Pat Alcock, Brenda Bishop, Sue Gregory, Mag Hayes, Jackie Roden, Sandra, Sharon Johnson, Kath Marsh, Dawn and Mary Wilcoxon, Nellie Asbury, Kath Knowles, Winnie Osborne, Wendy Armstrong.

Supporters: Local trade union branches, NUR workers in Paddington, Penguin Book Company workers, men at Treaton Tinsley Marshalling Yards, Ashfield East Ward Labour Party.

BILSTHORPE WOMEN'S SUPPORT GROUP

''Our group didn't get going until May, we had a job to find anywhere for a kitchen.'' Joan Preston highlights a problem which was by no means uncommon in Nottinghamshire. Bilsthorpe women had approached a local district councillor, a former miner, but he declined to help, saying that he wanted to stay out of the dispute, to remain 'impartial'.

> We tried the village hall, the Scout hall, the St. John's Ambulance Hall, no good, nobody wanted to know. We felt like outcasts. Ann Black, one of the women in the group, was on the committee of the Bilsthorpe Jazz Band, so in the end her influence got us the Applejacks Hall. (Most mining communities have a local jazz band, with young and adult musicians. These bands are major features of miners' rallies, festivals and parades, bringing not only music but bright swirls of colour as well.) We *got* it without too much trouble, all the bother came later.
>
> We had to accept certain conditions, it was only to be used for eating, no talking about the strike, no feeding or entertaining of pickets from elsewhere. We wouldn't have got it if we hadn't agreed, but how could we not talk about the strike? We never talked about much else, nothing mattered but the strike, that was why we were there. As time went on, they kept putting more and more limits on us. We had to stop doing early breakfasts and could only do

> dinners. Then we couldn't use it till 9.00 am and had to be out by 10.00; the afternoon limit was cut back from 5.00 pm to 2.00 pm. We knew why they were doing it, to try and spoil things for us, to make the men go back to work.
>
> Then they said we couldn't store food here; and they wanted all the big pans back, all just to make it hard for us to manage. But they never succeeded in driving us out, just made things more difficult.

The hall was not very homely, large and draughty with bare boards. The kitchen was tiny, far from adequate for the kind of cooking necessary, but there was a stove and a sink and the women brought their own utensils, crockery and cutlery. Janet Bone managed all the cooking, with other women helping, and Joan Preston was responsible for the weekly shopping at Rainworth Coop. Food parcels were packed up every week. There were about a hundred men out on strike at the beginning, but this number fell to thirty nine at the end.

Bilsthorpe was twinned with a group of people in Finsbury park, London and with a group of women in the Kings Cross area. Money from these groups came to about £120 each week and Newark Labour Party supporters sent £25, sometimes more, as well as clothing and food. The group had fewer visitors than some of the others as Bilsthorpe is off the main road and the Applejacks Hall is tucked away at the far end of the village. Jef and I went twice a week in the first seven months of the strike, then once a week, sometimes taking friends and supporters with us. We were always welcome, there was no ceremony, just "Are you all right?"

Joan Preston describes their regular trips to London, to speak to their supporters and have a short break:

> It wasn't like making a speech, standing on a platform. We sat around and told them what things were like, what was happening to us. My little lad suffers from cystic fibrosis. He has to be kept warm, he mustn't catch colds or his chest is very bad. They understood things like that, especially when we had no coal in the winter and the house got damp. It was like telling the personal things to friends.

Janet Bone takes up the story

> We found their way of living very different from ours, a bit shocking at first, I must admit, but they were good people, they gave us money when they hadn't much for themselves. We liked them and though they were very different from us we got on.

The women shared stories of the hardships that they each endured. Bilsthorpe women learned of the harassment, abuse and inequality in the way the London women were treated. New horizons opened up for both groups as the supporters learned about the coal villages and their life styles. Previously non-political, some never even having voted before, Joan, Janet, Liz, Kay, Janice and the others found a new awareness and saw many things in a new perspective. Though not intending to be politically active, they will not take so much on trust again. They tell me that I had some influence on them. Often, it is true, I challenged some of the chauvinistic attitudes, but it was all so good-humoured that I never really knew if the men were as outrageously 'macho' as they professed to be. I was not sure if they were trying to tease me out of the ideas I was presenting, or at least stop me from having any effect on the women who listened to them. I found that many of the miners responded in that way when conversation became uncomfortable for them to handle; they would turn it all into a joke rather than discuss it in a serious way. They would turn to my husband, expecting support, and would be nonplussed when they got none. They dealt with us finally, I think, by deciding that we were some more 'eccentric' friends they'd acquired during the dispute.

The women, however, were really interested, when they had time to sit down and talk about things. They backed me up when I chided the men for their attitudes to domestic responsibility and for seeing the role of women as subordinate. I encouraged the women to see how capable they really were, with their own points of view, well worth expressing. All this began months after I started to visit and when I knew that I was accepted. Even then I stepped cautiously, for I was a guest in their 'home' and new ideas were coming to them from so many quarters and making demands on the women which were hard to handle. What was important was that we respected each other's views and took from each other the ideas we wanted.

At Joan's house, the day after the strike ended, the women were depressed and angry. In tears, some of them felt that it should not have ended that way. Deperately tired, worried by increasing debts, not really knowing which way to turn, they still wanted to go on fighting. They knew that they had not finished.

Women Involved in the Group: Joan Preston, Janice Shaw, Liz Dully, Ann Black, Janet Bone, Janet Worker, Kay Stoneman, Kath Green, Ellen Monington, Janice Black, Lynda Calladine, Judy Johnson, Peggy Hall.
Supporters: A group in Finsbury Park, Kings Cross Women's Centre, Newark Labour Party, Mary and Fred Turner who brought eggs, poultry and home-made pies week after week for the duration of the strike.

BLIDWORTH ACTION GROUP

> Like the rest of Nottinghamshire, it was all very confusing here in the early part of the strike, with men going in to work but staying out when there was a picket. We had about 800 men out at first, but this went quickly down and down; then things were more settled until just before Christmas, when a few more went back. This is a mining village so most of our men live here, but a few come from as far away as Newark. We started to meet each other on the picket line at the pit, taking soup, drinking chocolate, potatoes to roast and coal for the brazier. We knew each other by sight and some of us were friends already.

It soon became obvious that there would be more to do than this and the women realised that they would have to find premises. There were pickets coming in from Derbyshire and they needed breakfast. When a young man offered the use of his house in Grange Road, which was empty, the women accepted it.

> You couldn't shift, it was too small for what we wanted, so we

decided to ask our local County Councillor, Les Jones, a miner, for help. He was on the management committee of the Youth Club and we thought we'd be well away in that building for the rest of the strike. He kept putting us off and 'thinking about it'. Finally we'd had enough and we decided to take it. The men didn't agree, but we marched up the road and forced our way in. It wasn't very pleasant but we were in and the men came up at night and joined us.

There were police already patrolling the village by April and they surrounded the Youth Club and asked everyone to leave. When the fifty occupants refused, they were threatened with arrest. By pleading, arguing, promising to be of good behaviour, to do no damage and cause no disturbance, they persuaded the police to leave them in occupation and go away. Pauline Radford declares:

> I don't know now how we had the cheek to do it. It was our first time demanding something and not being put off. We were all together and that gave us the guts to do it. And we learned a lesson that night, not to be put off. We needed that lesson for the rest of the strike, so it's a good thing it came early.

This early event set the pattern for their attitudes throughout the dispute. The men and women will never forget the night they spent in the Youth Club, supplied with sleeping bags, coffee and fish and chips. The women got the worst deal, sleeping on the cold floor in a big room with no heat, whilst the men were in the kitchen with the ovens all lit. Doreen Humber remembers:

> We couldn't get to sleep, we'd had enough, it was perishing cold so we moved in with the men and we all stayed there till three o'clock when they sent up supplies for breakfast from the house in Grange Road. The men were picketing at 4.30 am so we had an early breakfast. We hadn't slept much, any of us, we were too excited.

At 11.30 the management committee of the Youth Club held an emergency meeting and offered the use of the hall, but just for the women and children and not for the men. The men accepted but the women, flushed with one victory, refused. In the meantime the pit manager offered them the Village Hall. Betty Savage says,

> We were told we could feed everybody there but not 'foreigners'. We said we hadn't any, but we knew who he meant, Derbyshire or Yorkshire pickets. We fed everybody who needed feeding.

The place, known as 'The Glasshouse', became home for the strikers, women and children for the duration of the strike. Although Blidworth is rather off the beaten track it soon became the focus of attention for many who heard about the work being done there. Perhaps someone had heard Doreen Humber, Pauline Radford, Sue Petney or Yvonne Woodhead speak at a meeting, or had read of the police invasion of the village or seen these events on television. Journalists, writers, film-makers, solicitors and supporters arrived day after day to get first hand information about the police behaviour in the village. Women appeared on television and radio programmes, their photographs were printed on cards for fund-raising material, their stories appeared in left wing papers and pamphlets.

The women never tired of repeating the story of the police invasion of their homes and streets, or of the arrests which took place in the village. Doreen, particularly, became involved with solicitors and barristers from London who helped to prepare cases and defend the strikers. Gareth Pierce, Louise Christian and Marguerite Russell were not only legal advisors but became good friends.

Blidworth women showed early initiative in getting out and about establishing supportive links. Paul Thompson from Nottingham became their first teacher, helping them to organise and conduct meetings, keep minutes, prepare an agenda. Doug Shaw, Maggie Green, Keith Toms, Jim and Sarah Ryan, Clare Mayers, Alison New, Nicky Clegg, Sue Maddock and Mr. and Mrs. Thornton are all people with whom the women of Blidworth had close contact and who became their friends. The family grew and grew as the strike progressed.

Whenever Jef and I arrived, as we did twice a week for the most part, there would be visitors. They would be readily absorbed into the work, serving food, washing up, going down to the local picket line, listening to stories of personal involvement. Lyn Beaton would often be there, getting background material fpr the book she was writing about Doreen Humber and Pauline Radford. The activists

Top: **The kitchen at Blidworth; Left to Right: Ann Greaves, Elaine Taylor, Sylvia Brown. Bottom: Sylvia Brown and Elaine Taylor.**

would be fully occupied with bank books, accounts, orders for food, checking and arranging stocks, preparing speeches, answering supporters' letters. So successful was the group that they estimate that in all they handled £80,000. This paid for food, coal, cars and telephones for pickets and 'key' workers, electricity 'holding' payments, gas bills, buses for rallies and meetings, holidays, parties and Christmas presents.

The women surprised themselves by their achievements. They were amazed at how much they learned and how much there was to learn. The boundaries of this small mining community were pushed back to include experiences beyond their imaginings.

> Our world got bigger—professionals, feminists, punks, lesbians and gays, black people and Asians, we met them all and we stopped judging people by how they looked or how they spoke, we're all people.

Like all the groups, they had problems. One of the worst was the drift back to work and with it the loss of male comrades and women who had worked so hard in the first seven to nine months of the strike. There was much bitterness and blame, a sense of loss and a great sadness. Often in addition there were niggling problems which, when anxieties were heavy and tempers frayed, assumed huge proportions.

Relationships with the men of the strike committee were good, though the women insist that they controlled their relationship with the men. Sue Petney describes how the women came to see themselves in a new light, and how surprised the men were to see what they could do.

> We could do anything, for ourselves. Really the men are like cockerels with bright feathers, strutting about, but at heart they're only little boys. We've really got the strength, only we don't realise it. Women don't think they are as capable as men, when often they are more so.

In the final months of the dispute, when the end seemed inevitable, the women were already wondering what they would do after the strike.

One thing we have to do is show the scabs that we don't only work for ourselves, we've got to show them that we work for the community, so we must do things after the strike to make the village a better place for everyone. We want to get onto the community centre committee and work especially for the kids. There's nothing for them here. We saw other places in the strike and we want what they have for the kids.

A QUESTION OF LOYALTY

by Sue Petney

Why do you cross the picket line?
Answer me every one.
I've searched for the reason,
But answers have I none.
Will you keep on crossing the picket lines
Till all the pits are gone?
There are people on this picket line
Who have thought their reasons through,
They are fighting for a future,
For me, for themselves and you.
So why do you cross the picket line?
Is your mortgage payment due?
I'm only asking questions,
The answer's up to you.
There are men who've stood on picket lines,
They gave their all and died.
We've lost only money,
But you have lost your pride.
So why keep crossing picket lines?
Valid answers have you none.
Will you keep on crossing picket lines
Till all the pits are gone?

Women Involved in the Group:

Committee:

Chairperson:	Doreen Humber
Secretary:	Annette Holroyd, then Chris Tucker

Treasurers:	Annette Holroyd, Margaret Groves, Mary Dickey, Pauline Radford
Minutes Secretary:	Sue Petney
Welfare:	Ann Greaves
Members:	Yvonne Woodhead, Betty Savage, Ann Penny, Maureen North, Glynis Gitcham, Elaine Taylor, Elizabeth Anderson, Betty Sturtivant, Betty Rowbotham, Yvonne Dorsey, Pauline Howarth, Chris Tucker.

Other Women: Marilyn Strang, Jane Smith, Wendy Mainwaring, Ann Roberts, Sylvia Brown, Janet Sheldon, Denise Crossland.
Supporters: Sneinton Labour Party, Bill and Carol Simpson from Sherwood, pensioners Frank Leper, Mr. Haman, Ernie Hatton and Bill Smith, Cambridge, Harrow, Burton Joyce Labour Party, East Harrow Labour Party, London Borough of Camden, Housman's Bookshop in London, NGA in Essex, NUJ in London.

BOLSOVER WOMEN'S ACTION GROUP

Although Bolsover is a Derbyshire town, the pit is in the Nottinghamshire Area coalfield and for that reason the support group was invited to be part of the Nottinghamshire central group.

Toni Bennett explains that there was already a group of women who had come together for canvassing for Tony Benn in the election campaign in Chesterfield. The group had stayed together to campaign for the unemployed and other causes in the area.

> We had a ready-made organisation when the strike came and I got the task of organising in Bolsover, so I was part of the Derbyshire Women's Group. We got a hundred food parcels sent out to us for the start of the work. Then I heard that Joyce and Carol Clarke were already planning a group so it seemed like common sense to get together. We started to arrange fund-raising and local collections. There was not much support locally, though the local vicar, Reverend Britton, was good, he let us use the church hall to pack our food parcels and for jumble sales. The first few weeks were

a nightmare, with 400 parcels needed and only enough money for 100. At first we could only give parcels worth £2 and people got one every three weeks. We gave priority to single men living alone, and to families with children under school age. Then, when we got more money, we could do better.

The local Town Council was unable to give a donation from the penny rate as they would have been surcharged by the auditor. Instead they made a donation of £50 to the District Council Chairman's Appeal; as a result of this appeal, food parcels were distributed throughout the whole of Bolsover District. Unfortunately, the appeal only raised £1,000 so parcels were only given out three times under this scheme. Money was always short. The main problem was that Bolsover NUM members were raising money around the country for their strike centre and though food and money were coming in, they decided it should only go to miners and their families from the Bolsover colliery. People living in the town, but normally working at other pits, were excluded. The women tried without success to get a better arrangement.

We should have been able to work together, to share responsibilities, we were very disappointed that this did not happen. The NUM Strike Committee had a canteen at the Assembly Rooms but this was for pickets only. They had breakfast and mid-day lunch, then sandwiches, then just breakfast for which they charged 20p. We never had a kitchen and I envied the women in the other groups, working together.

Toni went out speaking to raise money for both Chesterfield and the Nottinghamshire central group. It was not possible to go out fund-raising for Bolsover because the men were doing this. When the group took money from Chesterfield, they did not take any from the Nottinghamshire group, and vice versa. Although they were beset with problems, they tried to be cheerful and resilient, working as a small team, enjoying jokes, marches and rallies. They picketed regularly and Toni was arrested for 'a breach of the peace' and 'using threatening words and behaviour.'

Coming from London before she married and moved to Bolsover,

Toni was used to seeing police violence. She particularly remembers their treatment of black people. Sue Mitchell will never forget or forgive the way in which her husband was treated by the police.

> They threatened to take a crowbar to the floorboards of our house when they accused him of being in possession of stolen goods. They said he'd broken a window, slashed car tyres, broken a gate—all false. He'd done none of those things. They haven't stopped harassing him since the strike ended, either. They are determined to get him one way or another. They never charged him, how could they? It was to intimidate him because he was on strike.

The deep hatred expressed by Sue is felt by many people who respected and trusted the police before the strike.

Women Involved in the Group:

Chairperson:	Ann Cutts
Secretary:	Toni Bennett
Treasurer:	Denise Porteus
Organiser:	Joyce Williams

Other Women: Carol Clarke, Sue Mitchell, Doris Floyd, Dot White.
Supporters: Mark Hogg, aged 11 years; a 90 year old socialist; Dennis Skinner, Tony Benn, Nottinghamshire Women's Support Group.

CALVERTON WOMEN'S SUPPORT GROUP

This group originally worked to help 120 families dependent on the pit at Calverton and one family from the Gedling Colliery. By the end of the strike the number had fallen to 39.

Maureen Hutchinson, Audrey and Rita Abbott already knew each other slightly when they got together to organise in an area where there was very little support except from some retired miners. Rita Abbott had been involved in earlier miners' strikes.

> There were no premises for us in the village so we had to concentrate on giving the best food parcels we could. I knew what had happened

> in the 1972 and 1974 strikes so I was prepared, only I didn't think it would last so long. It was very helpful to have the central group. We went to a meeting in Mansfield and we came back with £25 and lots of advice, especially about getting out of Notts and looking for support. It was obvious we weren't going to get much help here.

Rita described how the women went out to raise money.

> We used to feel terrible, sick headaches and stomach pains, but we kept going, we had to. We even went to Holland. Women took things to sell and auction, things from the strike. Everybody wanted to know what was happening, especially about the women and what we thought of Arthur Scargill.
>
> We used to go picketing, regulars we were, locally and all over. We went on all the rallies and marches, it was great to go to the other pit areas, especially to Wales. We liked being with the other women. It made us all feel good. We liked having people to come and stay at our houses. We hadn't got much to offer them but what we had, they could have, and we had plenty of laughs. We all shared everything.

Somehow the women seem to concentrate on the good times, remembering, for example, the fun and the excitement they had when a tanker overturned. It was carrying slack coal to a nearby power station. Everyone rushed out to collect the precious fuel. It was dirty, back-breaking work getting it up and then, to their disgust, they found that it only burnt for a short time. It smouldered and then went out, leaving the grate full of ashes, which had to be cleaned out every half hour to get any sort of fire. All their hard work had been for nothing. Somehow they manage to laugh about it now. A really good time was when a torrential rainstorm washed potatoes out of the soil into a ditch. This was too good a chance to miss, "We all went down into the ditch—free potatoes, we lived on them. And we had some fun getting them out of the ditch."

There were bad times, too. Often, when funds dropped, the women knew there was no alternative but to reduce the value of food parcels, and this was bad for morale. It was things like this which would drive men back to work, and the women knew it. When they witnessed for themselves the behaviour of the police, their obvious bias towards the establishment and the scabs; when they saw the bullying of pickets, and when they experienced it for themselves, then they came to detest the police: "We've seen women dragged to the van cells by their hair. There was no need for it. We've seen men being knocked about for no reason, except for being there, for being on strike."

The women speak with pride of their children and their men, explaining, "It is a wonderful feeling, knowing you've got a man who's great and kids who didn't complain. That's worth being £2,000 in debt for." "And", says Audrey, "we're proud of ourselves in a way, for being strong. Some women couldn't do it."

Women Involved in the Group: Maureen Hutchinson, Audrey Mulholland, Rita Abbott, Andrea Swanwick.
Supporters: St. Ann's, Nottingham, Huntingdonshire and Bethnal Green Labour Parties, supporters in Sheffield and Forest Fields, a lecturer from Leicester Polytechnic.

CARLTON IN LINDRICK WOMEN'S SUPPORT GROUP

There is no longer a working pit here; it closed in 1968 and miners were transferred to Shireoaks, Manton, Maltby, Whitwell, Yorkshire Main, Fens Workshop and Dinnington. At the beginning of the dispute there were 88 Carlton men out on strike, of whom 20 were single. At the end there were 41; of these 11 were single. "We only began in July, to feed the children during the summer holidays", explains Joan Ogley.

> The men organised food parcels in this area. Terry Healey, Pete Naylor and Bill Brown did it all and we decided to get a group of women together specially to feed the children when they didn't get school dinners.

Raising money became the way of life for the group for the next nine months! Jumble sales, raffles, bingo, coffee mornings, all brought in a steady income; charges were very low, 5p for a cup of tea, 2p for a bingo ticket. Jennie Staples organised the sale of logs for 20p a bag, and the chopping was done at Joan's house by a team of men.

> As soon as our total income in the building society reached £274.75 we started buying food in bulk. We got Coop vouchers, through Manton, from the Nottinghamshire central group. Manton and Shireoaks men walked to London, specially to raise money for the children during the holidays, and we got a share of that, about £200. The County Council gave us the use of the Carlton Park School. They tried to persuade us to feed the children of the unemployed, but we refused. We pointed out that we were getting stopped £15 from our benefit and they weren't. Altogether we raised £578 for that school holiday.

Barbara Slack, the school cook, Edna Dodd, cook at Crown Street School and kitchen staff members Jean Brumpton, Margaret Parkin, Pauline Williams and Ann Icliffe all gave their help voluntarily for the whole period. About 86 children came for a mid-day meal and they had fruit, sweets and chocolates for the weekend.

The group continued to raise money and at the end of the school holiday there was still £648 in the fund; the County Council had, unexpectedly, given a considerable donation for the meals.

The meals were provided again when October half-term came, 60 this time. Money was still being raised and so it was decided to share out money for Christmas. Every family received £10 and every single lad, £5. There were wonderful parties with extra food parcels and presents.

As the strike went on into 1985, the women worked to raise money for the mid-February holiday, wondering how many more holidays would come and go. Money came again from Nottinghamshire County Council and what was left over was used to provide logs, so desperately needed. The women reckon that over a thousand bags were provided.

Unlike other groups, this group raised all its money locally, and only about five women did all the work. It was not exciting, but it was rewarding.

> It took our minds off the worries, about the hardship, the bitterness, the police and the NCB.

★

> It brought us more into the forefront of the strike, we began to go on marches and rallies and meet all the others, and we made our own banner. We were very proud to carry it, showing we came from Nottinghamshire and that we were still together.

The women remember all the letters they wrote at Christmas time, to the *Daily Mirror*, to Cadbury's, to Matchbox Toys, asking for donations for the children. They received lovely letters in reply, all regretting that they could not help. Another disappointment was a fund-raising fete, organised in August, in the Church Hall. It was pouring with rain as women arrived, carrying home baking, garden produce, cakes and biscuits, food for the barbecue, bric-a-brac to sell....and the hall was still locked. What a start to the day! Wet through, they found the caretaker singing the Lord's praises at a choir practice. The women had been forgotten. Only 20 people came and they made just £36.79.

> We had so many disappointments, but we had to keep going. We worked as a team and that was the best part. I think we had a grand time when you take everything all together.

Women Involved in the Group: Joan Ogley, Sandra Alder, Jenny Staples, Edna Dodds, Ann Icliffe, Stephanie Whittle, Julie Preston, Dolores Middleton, Sylvia Goodhall.
Local Supporters: The pop man, the baker, two ice-cream men, Joe Deaken, Watmore, Betty the Fruit, Mrs. Austin, Marlene Myers, Nora Leyton, Mary Lerigo, Mr. and Mrs Mitchell, Henry Illiffe, David Ogley, Jim Watson, Pip Adler, Jackie Staples, Brian Ogley, Tony Finney, Graham Coe, Pat Keeley, Joan Hoyle, Valerie Bochenski, Mrs. Pickerskill, Mrs. Farrish, Joe Ashton, Mr. Hazelhurst.

CHURCH WARSOP WOMEN'S ACTION GROUP

Church Warsop is in Nottinghamshire, but Warsop Main is a Derbyshire Area pit. Men living in the village were transferred there when pits in Holmewood, Clapwell and Pleasley closed down. Miners who live here are also employed at Mansfield and Mansfield Woodhouse.

The group was set up to help all miners living in Warsop who were on strike. As this was a Derbyshire pit, there were no problems of finding premises; all the Derbyshire men came out at the start. Lynne Dennet discussed the need for an action group with her husband.

> I had a baby of 4 months and twins aged 5 so I wasn't sure how much I was going to be able to do. It was hard to get other young women with children to come and join a group but finally we got twelve to come to Karen Booth's house and I chaired my first meeting. I'd never done anything like that before but it seemed to go all right and we organised some leaflets; we wrote them all, inviting other women to another meeting, open to everyone this time. Union delegates from Warsop Main came. They thought they were going to tell us how to do things. I think they thought they were going to control us. But *we* had invited *them,* we weren't having them tell us how to go on. They could help us, we weren't shutting them out, but we wanted to do things for ourselves.

During the first weeks of the group's existence, only about ten people came each week to collect food parcels. Alison Lovely explains,

> I think many people thought they could manage, they thought the strike would soon be over. A lot of people were too proud or embarrassed. But after two or three months things changed. Money very soon started to run out. Pantries were empty. We were doing 450 parcels every week before long, but even then some folks wouldn't come for one. We needed a lot of money to do all this, and we needed it every week. We decided to do a door to door collection, to people who hadn't got anybody on strike. But someone told the police and they stopped us.

Betty Heathfield heard of their plight and came to
with money and food from the Chesterfield Women'
However, this was still inadequate and women decic
alternative but to go out into other areas to raise n

were made with sympathisers in London and women went to the Glastonbury peace and rock festival where they collected £463 in two days. People there proved to be eager to help and this gave the women from Warsop great encouragement.

There were plenty of helpers to make up parcels. Some would come in and ask for their 'hampers'. With her ready humour Lynne Dennet would remind them, "This isn't Harrods you know, we don't do hampers, only carrier bags." Work finished, relaxing over a cup of tea, the women would always have one late-comer arrive asking, "Aren't you doing parcels today?"

In May, when the children were on holiday from school, it was decided to open a kitchen and the Community Hall was made available. After this it was felt that it would be good to have a kitchen for everyone to use and so rooms were given in the British Legion Hall. All the floors and walls had to be scrubbed before it was ready to use, and camping stoves and an electric boiler were acquired. Between 200 and 300 meals were served there daily. Before Welbeck had a kitchen, miners from that pit came for meals and their group donated money to pay for them. Unfortunately no fresh food could be provided because equipment was so basic.

Still there were people who did not come for parcels or for a meal, feeling that there were others in greater need. During August, Nottinghamshire County Council gave money for children's meals. Food was cooked in the local school and carried in relays to the British Legion Hall where about 300 children were fed. During this period there was great solidarity amongst the strikers, women and children. All the times when they were together were good:

> Eating together was marvellous for us all, even though it was really hard work. Going out together was smashing, we went to Skegness, paid for by the Union. We'll never forget that as long as we live. But some of the families didn't come to anything. They really missed out, they were isolated, especially the women.

Lynne Dennet believed that the strike would be a long one.

> I expected it to be about a year. I went out speaking all over, and abroad. I went to Holland with other women, to the National

> Housewives Union and to meet people in Limburg. At just two meetings we collected £5,000. So many Dutch folks have been to stay at our house that they call it the Dutch Embassy. Other women, Debbie Bell and Alison Lovely, went with me to Amsterdam to meet the support group there and in five meetings we raised thousands of pounds; £1,000 came to Warsop.

The Dutch people gave not only money, but also holidays for adults and children.

Alison recalls being on Felixtowe docks waiting to board a ferry to speak to members of the National Union of Seamen.

> A security man came and asked me what I was doing. I'd been to the toilet, I was really nervous about talking to all those men. He didn't believe me when I told him I had to go to a meeting of officials. I had to beg him to let me back into the building, he did in the end when somebody came looking for me.

They still giggle when they remember a meeting held in Glossop, in the local Magistrate's Court.

> There were some very good speakers on the platform, we were waiting our turn, thinking we weren't as good as them. Somebody on the front row was nodding with every word, we didn't know whether he agreed or whether he had a tic or something. We couldn't stop looking at him and we were beginning to giggle. Then the chairman slipped and fell off the bench, chair and the lot. And a man stood up in the audience and started to shout and get excited, on our side, he was trying to start a revolution perhaps.

By this time Debbie and Alison were near to hysterics, trying to control their hilarity and trying to hide out of sight, under the bench. Lynne tried to remind them of the serious nature of the meeting, but in vain, for by that time the whole audience was laughing and the room was in uproar.

Raising money was a never-ending anxiety. About £800 was needed every week to meet the cost of food parcels. Women were worried to the point of desperation and the kitchen was saved only by emergency funds from the Nottinghamshire central group and food from the Chesterfield Women's Action Group.

The return to work reached its highest level in November and by the end only 25 families were left with men out on strike. On reflection the women say that this year was a period of education undreamed of before the strike. They have seen socialism in action in the grounds-well of support from outside the pit areas and in the cooperation of people in struggle.

Alison Lovely was a Conservative voter from a Conservative family, brought up to believe that strikes were reprehensible. Her political affiliation is now to the Labour Party. Her experiences have brought about an understanding of the nature of industrial relationships in this country and the attack on trade unions. She hopes to take further study, possibly at Northern College, to develop her new insights to a much further extent. Lynne was aready politically aware but has had her experiences broadened and her perceptions sharpened. She looks back on the strike and describes events with a mischievous sense of humour:

> When bad things happened, I just wrote a poem, like the time when somebody smashed eggs all over our front windows, what a waste of eggs! When I went to speak at meetings I felt like some sort of pet, on show, a woman from the pit villages. I surprised people, I don't think they expected us to be so intelligent. We didn't only tell them about feeding the kids, we told them about the union and what the Coal Board was doing, we could tell them about the industry. We knew about politics, about other people suffering from the government, and folks abroad, they didn't think we knew so much.

Alison thinks the children benefitted from seeing so many new people and going out of the village to new places they would never have seen.

> And it was good for them to spend so much time with their fathers. Good for dads as well, they got to know their children, they never had the chance before. If they were old enough they learned about the dispute and about standing up for principles.

Lynne adds that no women left the action group because their men had gone back to work; they are all very proud of that fact.

THE PAT ON THE HEAD

by Lynne Dennet (written for Liz Hollis)

Here she is, come and see her
The pet miner's wife, she's over there.
No, not her, the one in red,
The one that looks underfed.

It's amazing, isn't it,
The way she's so articulate?
She really knows what she's about.
It's not just grunt, scream and shout.

I've heard she can do joined up writing,
Don't you think that's exciting?
Do you think she can read as well?
Ooh, I don't know, you never can tell.

Well, it's been an eye opener coming here tonight,
You know, this miner's wife is quite bright.
I didn't pat her on the head,
She patted me on mine instead.

THE EGGY PLUNDER

by Lynne Dennet

Middle of the night, they must have crept
As we all silently slept.
Splat! It hit right on target,
The kind they throw at beloved Margaret.
An egg, yolk dripping, down my window.
Thank God it were that they decided to throw.
I've cleaned it off before it were dry.
I bet they thought it would make me cry,
But I laughed instead,
It didn't quite fill me with dread.
But now it's made me wonder,
Who did this eggy plunder?

Women Involved in the Group:

Committee:

Chairperson:	Lynne Dennet
Vice Chairperson:	Glenys Morton
Secretary:	Karen Booth, then Angie Brewin
Treasurers:	Karen Langton, then Marie Maddison, then Debbie Bell and Chris Wheatley
Members:	Bev Alsop, Joan Wheatley, Tina Copeutt, Ann Brown, Alison Lovely, Debbie Bell, Pat Widdowson, Gale Chapman, Shelley Crowther.

Other Women: Janice Pett, Sue Morgan, Mary Maddock, Maureen Jeffries, Wendy Griffiths, Madge Crow, Elsie Potts, Vera pringle, Lil Ward, Pat Smith, Tracey Adams, Kath Cantrell, Sue Fairfield.

Supporters: Belper, Glossop and Matlock Labour Party members, Tony Benn and Denis Skinner, Bob Pringle, Cyril Ward, Rena and Michell of Amsterdam, Hank and Will of Utrecht, artists Joolz, Nik, Rik and Justin, New Model Army Group, Langwith Action Group members, especially Pat Crewe and Rita Hill.

CLIPSTONE WOMEN'S SUPPORT GROUP

The majority of people who live in this village depend on the pit for their livelihood. As well as Clipstone pit, men living there work at Thoresby, Crown Farm, Mansfield, Sherwood and Bilsthorpe.

As soon as the men began to picket here we took food to them. But we soon realised that that wasn't going to be enough. We made contact with Ida Hackett and asked for help and advice. We got the Community Centre for premises, but only for women and children. We decided it was better than nothing but we really wanted a place for all of us.

Jane relates the story of the next step.

> The plan was to stage a sit-in at the Youth Club. Somebody pretended there'd been a watch lost the night before and we went to see the caretaker. At about ten o'clock, as soon as the hall was opened for them to look for the watch, about fifty people arrived with sleeping bags and food. Everyone was feeling excited but worried as well, not knowing what would happen. Nobody had ever done anything like that before. It was frightening when the police arrived, but they didn't even try to get us out. All the time of the occupation, six days, the police helped, fetching bread for us. Everybody got to know each other and made plans for things to be done. Even the Yorkshire pickets came in and we got to know them as well.

In the meantime Isobel Duncan began negotiations to obtain another hall because, despite the occupation, it was obvious that they were not going to be allowed to stay in the Youth Club.

> The St. John's Ambulance Hall was offered to us. We took it although it was a terrible place and in a dreadful state, nearly derelict. When we went to look at it a dog with us turned away, in disgust I think. But it was all we'd got and we had to make it decent. The manager from Clipstone pit got us better toilets fixed. It was scrubbed from floor to ceiling and then painted to make it look more like home. We managed to get a cooker and plenty of chairs and tables. When it was all ready and we were doing meals, hardly anyone came. It was disappointing, but when their savings had gone, they came then. Some folks came for the fun as well as the food; we had games of football, like a community centre. Soon we had so many people—200 to 300 every day—that we had to have three sittings.

The work became very heavy, for in addition to the large number of meals every day, parcels had to be packed and delivered. As the work load increased, more and more helpers came. Among the funniest things they remember was pumpkin pie:

> Somebody won the pumpkin in a raffle or something. Lucky for us we had Claudia Jeffries, who's American, and she had a book

with a recipe in. It took three hours to cut it up, there were ten trays of it in the end. We kept wondering if it would be worth it, if anyone would eat it. It was a bit like egg custard and after a while people ate it, some even wanted a second helping. We made a good stew one day. We put everything in we could find, even picalilli. People asked us to do it again, but we couldn't cos we couldn't remember what we'd put in it! When the rhubarb was ripe in all the gardens we seemed to have that every day. The men kept asking for extra toilet rolls in their parcels.

★

The kitchen was a great place. I'm sorry for the scabs and their families, they'll never know what they missed. One day a little girl stood at the door and when we said she could come in she said, "I can't, my dad's a scab." The cookers were always over-loaded. I can remember when the fuse used to blow. We had to run through the streets to finish off half-cooked stuff on our cookers at home. It was funny that, carrying pans through the street, everybody looking.

Women enjoyed visitors to the centre and they were interviewed and filmed by reporters from Canada and Germany. Pickets were always welcome and always enjoyed the cooking. Some support came from local people, especially old age pensioners, but the women had to take the campaign out of the County. Miners made early contact with NUPE branches and the National Union of Seamen. Support came from Pendle and Whitby Labour Parties. A hotel owner in Manchester who had helped the Jarrow hunger marchers read an article about Clipstone in *The Guardian*. He appeared at the kitchen one day with his car full of meat, groceries and vegetables. At Christmas he supplied everything needed for Christmas Dinner, including all the turkeys.

Margaret Hood went to Belgium with a party of women to speak on the strike and to raise funds. Jane Holness also spoke at meetings and when money was short the group knew they could always turn to the central group for help.

All of us in the strike have changed. It has shown us new things, surprised us a lot; especially the police. We trusted them at first,

Scarborough, N. Yorks.

Enclosed is £1 note for your Solidarity fund, for with all the success for Victory for the Miners,
I was in the 1920 strike & was out for the 26 weeks, as you well know how we were legged down by ~~by~~ H.G. Thomas, railway leader, after the 14 day general strike.
Well I have relations at Rufford at that time, I havin't seen for years, I have forgot the last name, but my cousin, name Ruth lived in street at side of the Rufford Hotel, Well excuse short letter, also the writing as I can't hold the pen tight through an accident in the pit.
Yours faithfully

Brenda Prince/Format

Clipstone Women's Support Group
Left to Right: Back Row: Ann Campbell, Barbara Neale, Mavis Giles, Elsie Lowe, Jeanette Parker, Jodie Holness, Claudia Jeffries, Barbara, Margaret Anderson
Front Row: Catherine Carmody, Jackie MacSherry, Carol Potter, Theresa Strachen, Margaret Hood

> when we were in the Youth Club, they helped us, but they soon changed and so did we. They were violent, they tried to intimidate us, they were arrogant and did what they liked. We couldn't get justice, no way. We were on strike and that was a crime. What did they expect us to do, just take it all?

Even the children learned lessons in the strike. One man took his fourteen year old daughter on the picket line with him in the early morning before she went to school, to let her see and feel things for herself:

> I heard a little fellow three years old shout out to a lorry driver carrying scab coal, "Hey, you, that's my job you're taking". Those kids will understand even better when they get a bit older.

Sadly, the ramshackle old hut that was the Centre has now been demolished. There is no standing memorial to their struggle, but in their memories there remains much more than they can ever fully express.

Women Involved in the Group:
Chairperson: Ann Campbell
Secretary: Isabel Duncan
Treasurer: Irene Ripley
Other Women: Jean Bingley, Theresa Strachen, Elsie Lowe, Margaret Hood, Margaret Anderson, Jane Holness, Carol Potter, Barbara Neale, Nellie Moore, Mavis Giles, Helen Cornell, Claudia Jeffries, Sandra Haddon, Cath Carmody, Jackie McSherry, Vera Taylor, Jeanette Parker.

COTGRAVE WOMEN'S SUPPORT GROUP

Jennie Robinson called a meeting of women on 21st. April, 1984, passing messages to women through men on the picket line at Cotgrave Pit. Around 24 women came and they decided to form a self-help group for families living in the village who were on strike. Jennie contacted Betty Heathfield for help and was invited to a meeting at Ollerton Strike Centre.

> Ida Hackett was there as well and we got lots of ideas from other women; some had already started, so that helped. Then four of us went to the Labour Club in Chesterfield where all the delegates from the Derbyshire support groups came, it really helped us.

In the beginning, when about 100 men were out, local shopkeepers were keen to make donations, but then, when they realised that the strikers were in a minority, the implications for their trade quickly appeared and they withdrew support.

At first, Jenny's spare bedroom was used for packing food parcels, but when she got a job Margaret Brown offered her garage instead. No premises were available for a kitchen. In any case, there were not enough women in the group to manage one.

> Many women work at the Saxondale Hospital, it's close to here. Most are on shift work, so a lot of work fell on Margaret, who didn't go out to work and didn't have small children. Margaret Cool also did a lot, even though she had a morning job.

Denise explained that many women whose husbands were on strike got jobs to help out with the family budget. Women at work helped when they could, but some had young children and a job, so it was difficult for them to do as much as they would have liked. There was a steady group of about twelve women altogether and they became firm friends. Margaret Brown was the central figure and as well as working to provide the parcels every Thursday, she went out speaking for the central group.

> Visitors always stayed at my home, from Brixton and all over London. We had letters from supporters like NUJ members and Electrolux in Luton. If we'd had a kitchen like other groups, we'd have had more visitors.

"Most of us couldn't go picketing, we didn't have time or

Dear Comrade,

In response to your letter in the Morning Star 8-5-84.

I have pleasure in enclosing a cheque for £10.00 (ten pounds) to help in your struggle.

I remember as a laddie going to the soup kitchens in Durham, & being locked out, in the first year of my working life. How hard was the struggle then, however, I have always been proud to come from mining stock, & of being active all my life in the struggles of the working class.

May unity be born out of this struggle, & victory to you all.

Yours fraternally,

W Siddle

The kitchen at Cotgrave:
Eileen Tucker, Denise Palmer, Jennifer Chewings, Susan Dickson

transport.'' Jennie went to the women's pickets whenever she could and everyone tried to make trips to Yorkshire, as a family, to renew their sometimes flagging spirits and to allay the feelings of isolation in Nottinghamshire. ''Rallies and marches and socials were good, feeling together, it kept our spirits up, made us feel solid for the strike.''

Doreen Gladwin, a local Labour Party activist, helped the women analyse and interpret their experiences and make connections between their ideas and the philosophy of socialism. She worked with them and went with them to the stall in Nottingham where women collected every Saturday.

Denise Wright sees the strike as having a tremendous impact on the women who participated in all the activities.

> Old friends say they see big changes in us. They sometimes find the changes hard to accept. We're confident in ourselves, we say our piece. We know about politics, we know what's going on. They get uncomfortable, they can't join in, we know too much. We never did before, we are different now. We feel sorry for them because they didn't have what we had. They haven't wakened up yet.

There was a steady flow of cash and at no time was there any anxiety that funds would fail to meet their needs.

Women Involved in the Group: Eileen Tucker, Jennifer Chewings, Margaret Richardson, Denise Wright, Susan Dickson, Jane Brae, Doreen Gladwin, Peggy Hogan, Ruby Robertson, Agnes Wuinn, Angela Cool, Janet Robinson, Lynn Martin, Yvonne Bowers, Jean Bunting, Kay Bunting, Joan Hepple, Freda, Brenda Rutherland, Janice Sinfield, Edith Tuskin, Lynda Bexon, Tina Severns, Mrs. Bexon, Vera Wilkinson.

Supporters: A local butcher, the Bakers Union in Yorkshire, Mr. and Mrs. Brackenbury, London NUT members, Kingsway College, Watford, Scarborough and Rushcliffe Labour Parties, Nottingham Probation Officers and NALGO members.

CRESSWELL WOMEN'S ACTION GROUP

This group started work in April when there were about five hundred miners, half the total work force from the village, out on strike.

> Leslie and Elsie Marton and Lesley Brailsford started to do breakfasts and sandwiches for pickets at Cresswell Community House, which belongs to the Parish Council. They did it from their own money. We got the Community House with no bother because there were some councillors on strike. Then some more women joined and strikers went to London to collect money for parcels and food. We got a rota for cooking and we used to meet every Friday afternoon to plan for the next week.

Meals for everyone were first provided in June, with a very small cooker and two tiny rooms for dining. When the Summer holidays came, another group worked to feed children in the local school, helped by the domestic staff of the school kitchen. Problems which arose between the two groups were solved and a new committee formed with regular meetings to plan all the work and iron out any difficulties.

Celia Richardson remembers the fire at the Community Centre which destroyed everything, including stores of food:

> It was a terrible blow, and a shock, everything went. But we got a lot of publicity and a lot of help to get started again. We hadn't got anywhere to work so we asked the local church and they said we could have the Village Hall; but we had to pay £30 a week rent. It was a better place than we had before. We had two good cookers and a lot of space and the pickets could use it as well as women and children. But having to find £30 before we had anything to spend on food was awful, and we had to move everything out when somebody else wanted to use it. Johnny Bleakley, who was unemployed, let us put all our stuff in his front room, in his house near the hall. He helped us a lot, in the kitchen as well.

Pam Richardson describes another regular visitor to the kitchen:

> Johnny Crowder, a retired miner, he used to come every day for a cup of tea. He came in at about 9.30 am and stopped all day. He had two strokes, but he still kept coming, all through the strike. He was miserable when it closed, he couldn't understand why it wasn't open any more. Sad. His son worked all the way through.

There was almost no support locally for the strikers. However, the Methodist minister would have given use of his church premises had he been able to, and he gave occasional food parcels.

> It was rotten for the kids, not being able to go with the others on school outings. So we had to do something to make up. We had two great days out, at Butlins and at Drayton Manor Park. They saw a lot of what was happening on the picket line, it's near the school and they used to hang over the school wall. The teachers wouldn't let them wear strike badges at school, but one day they all came home with police badges, 'I've met the Met', it said on them. The police had been giving them out in the village. That's hypocrisy for you, when the kids had seen what the police do to strikers, like handcuffing two of them for hours, to a lamp-post outside the school; two young lads, handcuffed together like criminals, for everybody to see.

Pam describes their deep and unremitting hatred of police activity:

> We'll never forgive them for what they did to us. They had the nerve one day, two of them, to come to our kitchen and ask for a cup of coffee. We offered them a cup, for £5. That's what we think of them. We got one over on them though. They were chasing a young lad, a picket, and he rushed into the Community House. We locked the door while he stripped and got some clothes from the jumble. You could always find plenty of jumble clothes here. By the time the police forced their way in he was standing at the sink, peeling potatoes, calm as you like. They didn't recognise him. But we recognised one of the coppers, he was on the picket line later, dressed up as a striker, provoking the lads to violence, trying to get them going.

★

> When the men went away to other pits to picket, we went on the picket line here. We went to Whitwell and Cresswell. I've used language I've never used before, I can tell you. We felt so mad, we had to swear. They lifted men and they'd be gone for hours and hours. Nobody knew where they were. And when you phoned the police station they'd say they didn't know anybody by that name, they weren't there. All the time they knew, they'd got them there but they wanted to cause you worry.

Money was always short and much of their funding came from the Nottinghamshire central group. To supplement this, women raised money locally whenever they could and Joan Cole went to speak in London to raise money for the group. From September, anticipating the need for a good Christmas, money and special foodstuffs were put aside to make sure that the children, especially, had a good time.

As in all parts of Nottinghamshire, here in Cresswell families suffered from divisions between striking and non-striking members. Pam recalls with sadness that family events were postponed or disrupted by quarrels and bitterness:

> The vicar of Cresswell called a service just after the strike ended. he said he wanted to heal the community, the bitterness. We didn't go, we've still got splits in our own families, never mind the community. It will take a very long time to forget—if we ever do. I'll never forget the young strikers, always hungry. My two lads are single, they were on strike. I used to have other lads on strike at our house for a meal, they'd eat up every bit and any scraps, any fat that was left on the plates. You don't forget that.

Women Involved in the Group: Joan Cole, Pat Reynolds, Cath Chambers, Pam Richardson, Celia Richardson, Gail Cole, Angela Belfitt, Marion Stevens, Mrs. Sergeant, Vera Taylor, Peggy Harding, Doreen.

EDWINSTOWE WOMEN'S SUPPORT GROUP

In April, approximately two hundred men living in the village were on strike fron Bevercotes, Clipstone, Ollerton and Sherwood pits. Ann Brown, the wife of a retired miner, went one day to a picket line at Bevercotes to take biscuits, cobs and drinks to the six men on the official picket:

> I got caught up in a mass picket of Yorkshire miners and the police frightened me; not the pickets, the police. I decided then we'd have to do something for the people on strike and I went to see Ida Hackett to talk about setting up a kitchen.

Pam Wingate had already met a woman from Welbeck who told her about her work in a women's group and so Pam and Ann took some of the local strikers' wives to a meeting at Ollerton where they heard what others were doing.

> Then we called a meeting ourselves in the Miners' Welfare and we decided to try and get the old Welfare building for our kitchen, it was being used by a local playgroup and the Sons of Rest—they're retired miners. We'd heard about other groups having a 'sit-in' to get a place so we told them we'd do the same if we didn't get it. We said we'd occupy it. We won.

The old Welfare kitchen was made available for five days, from 9.30 till 11.00 am at first, then until 1.30 pm.

> We couldn't store food or clothes in the Welfare. At first we had to keep everything in Jessie Murphy's house, then later we used Malcolm Deeming's garage. We only had four hot plates and a warming oven but we managed to cook a hot meal every day and we took food to the nearest pit, Thoresby, for the 10.30 am picket. Whcn we had enough we packed a parcel worth about £3 to £4 every week; if not it was once a fortnight. If you wanted pet food you had to put £1 in a float for the kitchen fund.

The group was twinned with Islington Support Group and this brought in between £100 and £600 weekly. Dockers from Tilbury gave wonderful support, sending money every month and £1000 for Christmas. Ann talked about local collections:

> We used to go collecting in Nottingham. Then, in the summer we thought we'd go into Sherwood Forest, when all the visitors were there. We were doing well in the picnic area until an official came and made us stop. It wasn't easy collecting, some women didn't want to do it. A lot had young children so in our group it was the older women, like me, who did most of it, we were the active ones. I got abusive telephone calls. One said there was a bomb in my kitchen. I got sick of these calls and so one day when I picked up the phone, I said, "This is Ollerton Police Station". I didn't get any more. My window was smashed by a brick coming through, and down at the Senior Citizens Club the atmosphere got so bad I had to leave.

Women went often to the picket lines at Ollerton and Thoresby and will never forget the police violence they saw there.

> It was bad, really bad. The police could do just what they liked to the pickets. But it was good when they were looking for the Yorkshire pickets in Sherwood Forest. It was lovely weather then, they were sleeping out, we'd got their cars hidden away, in people's garages, away from the police. I can remember opening our door to let the pickets come in and hide when the police were chasing them all through the village. I had one in my kitchen. I shut the door in the policeman's face and I wouldn't let him in. I never thought I'd do that, or feel like I do now about them. I think we all do.

Ann is concerned by her own attitudes, having always been law-abiding and respectful of the police.

There were many changes in the Edwinstowe group. As men returned to work, their wives left and took no further part.

Women Involved in the Group:

Chairperson: Jessie Murphy
Secretaries: Elaine Murphy, Margaret McNee, Diane Beard, Pam Wingate
Treasurer: Ann Brown
Other Women: Diane Deeming, Mary Green, Mary Hallam, Eileen

Dorwood, Lynda Boden, Sheila Scotham, Liz Benson, Margaret Shay, Mavis Dixon, Marie Butler, Pauline Walker, Melanie Hallam, Maureen Smith.

GEDLING SUPPORT GROUP

Gedling is not a colliery village in the sense of having a large number of families living there who are dependent on the pit for their livelihood. It is not a close community of miners and their families. Gedling pit men live in places scattered over a wide area. For this reason the women had no kitchen and no central meeting place.

Four weeks into the strike, food parcels were being made up at Pam Kenny's house:

> We were buying food in the shops here and we didn't get any discount. I'd called a meeting at the Miners' Welfare and eleven women came. I didn't think that was many, but then at the second meeting we only had eight. Fran from Derby came to help us and she got notices printed to try and get more women, but we didn't get many. I wrote letters to local shops but they didn't want to know. We were getting about £6 to £40 from the local Communist Party and from £4 to £7 from the Militants. We were desperately short.

Alice Terry and Pam decided to go to the Cooperative Society Conference in Blackpool and collect there:

> We stood with our collecting tins in the pouring rain. There were two entrances but we were too scared to separate so we could cover only one door. We got £38, that's all, and we had to pay £20 for our petrol for the trip. We felt dead miserable.

Pam decided she wasn't coming to Blackpool without a bit of fun, so she had a donkey ride.

> You should just have seen me, wet and miserable on a wet and miserable donkey. And he smelt. Just me and that smelly donkey

I had a lucky win in a charity 100-club this week. Here is half the winnings for the fund. In the cosy South-East you have friends.

Yours fraternally

Sheffield, Yorks. June 1984

WITH THIS NOTE I HAVE ENCLOSED A £10 DONATION TO THE SOLIDARITY FUND WHICH I KNOW WILL BE PUT TO GOOD USE. I WISH YOU THE BEST OF LUCK IN THIS DISPUTE AND I REALISE YOU ARE FIGHTING FOR YOUR LIVELIHOODS. SO FAR DURING THIS DISPUTE I HAVE BEEN APALLED AT THE GROSSLY UNFAIR MEDIA COVERAGE AND AMAZED AT THE IGNORANCE OF SOME SO CALLED JOURNALISTS. I FOR ONE ARE NOT FOOLED BY IT. NEITHER ARE MANY OTHERS

BEST WISHES

West London

> on the beach, I must have been mad. But it was a bit of fun and it made us laugh though we didn't much feel like laughing.

Women from Cotgrave came to Gedling to offer help. This sharing of expertise and resources happened all over Nottinghamshire and was much appreciated, easing feelings of isolation. Joining the central group meant that regular donations of money, food vouchers and food were available. All this helped to maintain the regular provision of food for weekly parcels. Pam says,

> I can still recite the contents, it's in my memory for ever. I think it'll be on my heart when I die: beans, soup, sugar, corned beef, meat balls or Irish stew, cornflakes. That was until some of the men went back, then we had a joint of meat three times and other, better things in the parcels. But there were always basics.

There was no early twinning out of the area although help came from Luton Labour Party members. People from Luton came to fetch items to raffle, but then there was a strike of car workers and, inevitably, support fell away. Drake Labour Party in Brockley began to make regular collections and by September were able to send £40 to £100 every week. Gedling Labour Party did not collect the 50 pence levy but individuals made their own donations.

> We managed to find time to go out speaking. We never really liked it; it didn't seem to get any easier, we were always nervous. I went to London with Joan Sim. We were early and we found a park to sit in. I had a little bottle of brandy and I kept having a sip to calm my nerves. Joan was sniffing smelling salts cos she had a headache. If anybody had seen us, me like a dipso and her like a glue sniffer, what a pair!

Alice Terry went to a TASS meeting in Exeter, thinking it would be just a small gathering: "Just imagine, it was a proper conference, with cameras and reporters." As guest speaker Alice felt overcome, almost too nervous to speak. However, the words came somehow and audiences always appreciated hearing a woman speaking about the strike.

Most enjoyable was when the Yorshiremen came to picket at Gedling:

> We cried with joy to see them. We made tea and sandwiches, gave them meat pies and chips. We wanted to do everything for them, we were so pleased to see them. Everybody wanted them to stop at their house. We needed them to give us a boost, and for the men out at Gedling, to lift their morale. We used to go on the picket line when we could and we liked the women's pickets organised by the umbrella group. They were really good. You saw women you hadn't seen for a bit.
>
> ★
>
> Because we lived down here we were cut off. We didn't have a kitchen or anywhere where we could get together like some did. At Christmas, when the other groups could have parties, we had nowhere to hold them, but we did have some marvellous toys and presents. I don't know where they all came from. At the pit a young lad who had gone back to work organised a collection for the families still on strike. He made £251. I changed the money into co-op vouchers straight away, gave them to everyone on strike and sent the manager a receipt so he'd know the money had gone to the families.

Some children from Gedling went abroad for holidays in France and Holland and this helped parents to feel that these experiences to some extent compensated the children for the disruption in their every-day family lives. Pam declares,

> Those of us who stayed with the strike have learned a lot. The strike took us out of ourselves and showed us new places, new people and new ideas. Do you know, somebody keeps sending me books from Albania, you'd be surprised what I know about Albania now! I never did before.

Women Involved in the Group:

Chairperson:	Joan Sims
Secretary:	Denise Wood
Treasurer:	Pam Kenny

Other Women: Chris Salmon, Shirley Skinner, Pat Morrow, Mrs. Moore, Janice Starbuck, Joan Price, Fran from Derby, Alice Terry.

HARWORTH WOMEN'S SUPPORT GROUP

We called our first meetings of the group in each other's houses, then we managed to get the Working Men's Club. We had about 160 men out from Harworth and there were some others who lived here, from other pits. A woman had come down from Ollerton to a meeting on May 16th to tell us what they were doing. They were already going well there. We decided to start with food parcels.

Kath Potts' home was the centre for the food parcels and when the parcel distribution was running smoothly, the women decided to try to find premises for a kitchen. As in other parts of Nottinghamshire, Harworth strikers were refused the Parish Hall, church halls, the Miners' Welfare Hall and even the cricket pavilion.

When the Whitsuntide holidays came along the School Sports Centre was given for feeding children and then Nottinghamshire County Council allowed an extension. The kitchen was available from 10.00 am to 2.00 pm during school holidays, for everyone, adults and children, and in school time for snacks only, from 2.30 pm to 5.30 pm. When the women were providing full meals, it was not possible to give food parcels as well. However, once they had a kitchen, support increased and donations came in larger amounts.

If only we'd had a kitchen earlier on I think fewer men would have gone back. It would have made a difference, kept us all together, seeing each other every day. We noticed that morale got better once we got the kitchen. People saw their kids being well fed, that made a terrific difference.

Money was accepted from the central group funds when needed and support came from Bury St. Edmunds, Thetford and Gainsborough Labour Parties, but without the help given by Fleet Street supporters the group could never have survived. Bassetlaw Council provided food parcels for about thirteen weeks and this eased the burden on resources. The women pay a special tribute to George Birchby, "He helped us in lots of ways but especially with Social Services." Kath explained that they had to go on the streets collecting:

> We did well in Manchester, collected £400, but Nottingham and Worksop were very bad. We had permits, it was all legal, but the cops harassed and intimidated us. They tried to move us right away from places where there were people about. They tried to get us into fields. How could we collect there? Well, they didn't want us to get anything. And we used to get abuse, it was awful but we tried to ignore it.

Some of the women recall one specially bad incident,

> We were collecting in Nottingham and a woman came past, shouting abuse. She came back again and did the same. It's almost unbelievable but she did it five times, just to give us abuse. We just had to swallow our pride. If we'd done anything back the police would have shifted us. We needed the money. We had to put up with it but it made us mad.

Maureen Garner and Christine Brown were the main fund-raisers at meetings out of the county. Maureen says they were very nervous.

> We were really scared to death. We didn't stand up at meetings, we had to sit down and talk cos we didn't think our legs would hold up. But it was good in another way, quite exciting and everybody was nice and friendly and pleased to see us. We made a lot of friends. But the meetings weren't easy, there were always different people and we were always nervous. Every time.

Christine speaks for many of the group when she describes their establishment of their right as women to more personal freedom.

> We are bored with life in the village, not with our families, not that, but with so few opportunities for us here. Life is dull. We've seen so much and we want to go on like that, learning and changing our outlook I suppose. We are a group of women together. We speak our minds, we are honest with each other. We've shared so much

> hardship together this last year, we know each other really well and trust each other now. Not only that, we are more political now.

A very large number of women have moved, *en masse*, into the local Labour Party branch and Constituency Party and are taking office, determined to make policy and effect changes.

> The men here are encouraging us, they want us to show how capable we are in politics. Somebody's going to get shaken up by us women.

Christine Brown speaks with great anger about the police activity in the village and her personal battle with the National Coal Board:

> The Yorkshire miners who came to picket at Harworth were not violent men. They never touched private property though they were accused of it. They were chased through gardens and houses by the police. They were dragged from the picket lines, really brutally, and they were beaten after they were arrested; just for being there.

> I was sacked from my job as an office cleaner at the pit. I went back with the men on 5th March, but I was told I was sacked. They told all sorts of lies, they knew I was on strike like the men. But they sacked me. I'm still fighting to get my job back.

Women Involved in the Group:

Chairperson: Christine Brown
Secretary: Ann Yates, then Kath Potts
Treasurers: Maureen Garner
Other Women: Shirley Woods, Jean Lees, Jackie Smith, Nina Kingston, Pauline Thomas, Alena Wright, Lorraine Noy, Linda Franklin, Gillian Cameron, Chris Davies, Julie Gee, Carol Hunt, Ella Davidson, Olive Clarke, Janet Lyons, Jackie Topping, Linda Jones, Val Brown, Julie McLaughlin, Jean Smith, Sue Richards, Jean Garner, Sue Meadows, Pat Gratton, Pat McGarry, Doreen Richardson, Carol Stubbs, Wendy Barr, Jean Thomas, Janet Weber, Denise Brennan.

HUCKNALL AND LINBY WOMEN'S ACTION GROUP

Linby Colliery has a long tradition of militancy and a strong union, so it is not surprising that there were about 555 men out on strike at the beginning; Hucknall had about 76 strikers.

The women who formed the group for the two collieries were determined from the beginning that they would work together, positively, for the strike, doing whatever was necessary to sustain it; hence the deliberately chosen name, 'action group'. Denise Perry remembers how her little house was packed with about fifty women for the first meeting; they stood in the kitchen and the toilet, and even overflowed into the garden.

> I rang round to women who had men on strike and invited them to come to our house for a meeting. So many came, it was chaos, we were all wanting to talk at once and people couldn't hear what was being said. We'd not much chance of organising a meeting but we managed to elect some officers. I was one of the secretaries; later, when I was at work full-time, Sandra Creed took over.

Someone sent a 'secret' donation of £200 with a proviso that a proper constitution be drawn up and sent to the donor, together with a list of all the women who were involved. The women managed to acquire use of the local Labour Party rooms, but only for packing food parcels. The premises are tiny, damp and very decrepit, but they overcame all these problems. Little by little they encroached until they had installed a cooker and a refrigerator. Soon they were managing to provide cooked meals every day. Meetings of the strike committee were held there as well and the Labour Party found that it was often impossible to get the use of their own premises. Every Wednesday there was a meeting of the women in the action group, the men from Hucknall and the men from Linby. This 'tripartite' meeting ensured communication between the three groups and that they worked together for the benefit of everyone on strike. Sally remembers:

> We were doing so well, everyone was working in harmony, then someone tried to spoil relationships between us by telling the Labour councillors that we were having orgies in the Labour rooms. It was all lies but it really worried us that we might lose the rooms, I don't

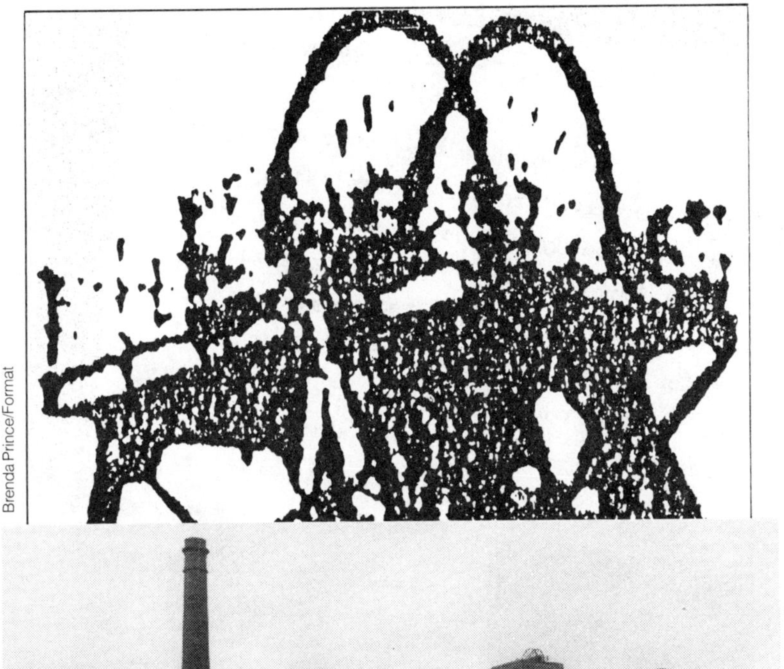

Brenda Prince/Format

Members of Hucknall & Linby Women's Action Group outside Linby Pit.
Left to Right: Louise Gower, Iris Wake, Linda Hickling, Sally Seabrooke, Rita Petcher, Dorothy Hickling, Pat Gower

Member, Electrical Power Engineers Association

John Lyons claims that the power engineers are not involved in the miners' struggle. This is clearly untrue. Apart from the general importance of the miners dispute to all trade unionists, EPEA members in particular can thank miners militancy for their own high standard of living. I enclose cheques for £500, which represents the value to me of the 1984 pay increase.

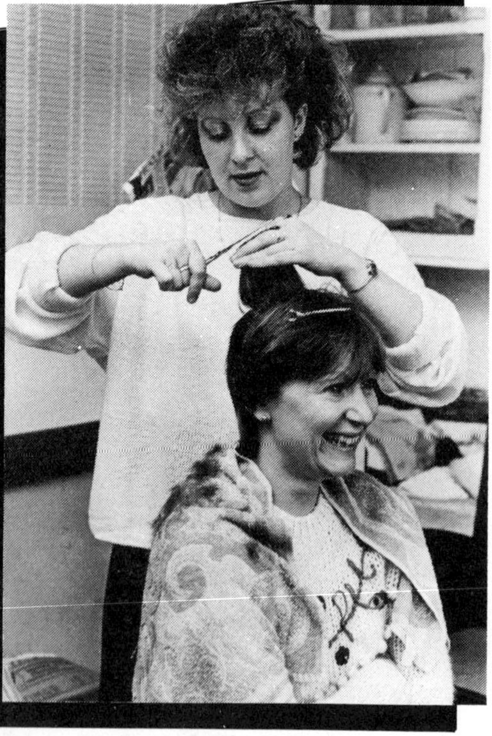

Free Haircuts for striking families in Hucknall

Brenda Prince/Format

Brenda Prince/Format

Striking family with Union banner, Linby Colliery

Brenda Prince/Format

Miners Kitchen, Linby

> know what we should have done. We had to have somewhere to cook and eat, and somewhere to meet. But nothing came of it, they knew it wasn't true, just somebody trying to get us out.

The women recount their attempts at street collections:

> We hoped we could collect for food on the streets. We went to Nottingham with the men, taking children as well, but everyone got moved on by the police.
>
> ★
>
> We had the Linby banner there. We wanted people to see who we were collecting for. There was a lot of trouble over boundaries and the police made us collect right away from where people lived or shopped or worked, though we did collect outside some factory gates.
>
> ★
>
> Black people were marvellous. They always helped. Sneinton market was a good place, that's a poor area of Nottingham. There's a large Asian population there.

However, it soon became clear that street collections which yielded only £4 in two and a half hours were largely a waste of time and it was necessary to take the campaign for funds outside the area. Links were made with people in Welwyn Garden City and in Brixton. Sandra Creed was one of the main speakers and in October Denise Perry went to Ohio, Pennsylvania and Virginia to raise money.

> The miners didn't even know the strike was still on, their television told them that it had collapsed three months before. They were furious about a film they'd made showing American miners in support of the miners in this country. The film had been cut and all the bits about support and solidarity had gone, so it looked the opposite, as if the American miners were not on our side. They really were mad. They were on strike themselves at the time, but even then they gave us $10,000 in two weeks.

American miners, both men and women, came to Hucknall to stay and see the truth for themselves, visiting kitchens, joining men and women on picket lines, taking photographs and hearing stories of the strike first hand.

Rita Petcher and Linda Lingard ran the Welfare Rights Centre, held every morning at the Labour Hall:

> People didn't know what they were entitled to, and when they did know, it was sometimes very difficult to get it. We helped people with forms, claims for free meals, clothing, everything that worried them. Danny Phillips from the Ollerton centre gave us advice and help when we needed it. It was very interesting work. Women were doing this all over in the support groups.

Dorothy Hickling, who was on strike from the pit canteen at Linby, supervised all the cooking along with Carol Roberts:

> We had some fun when the foreign food came. Nobody could read the labels, we had to go by what was in the picture, if there was one. We had some glass jars with no proper labels, you just had to guess what it was. Someone said they fed some to their dog and it almost died, but I think that was a joke. But even though they joked about it, some people liked it.

The women agree that the strike was very hard work but gave them great fun and satisfaction. Their horizons were expanded, their political education extended and deepened. Bonds of friendship were formed that will last for ever.

Women Involved in the Group:

Chairperson:	Sally Seabrooke
Secretaries:	Linda Hickling Denise Perry, then Sandra Creed
Treasurer:	Julie Parkin

Other Women: Colleen Walker, Cath Bakewell, Iris Wake, Nelly Butler, Val Betteridge, Dorothy Hickling, Carol Roberts, Rita Petcher, Linda Lingard, Chris Cain, Sharon Dennis, Gill Davies, Chris Straw, Marylin Stonely, Kath Jacques, Lyn Nixon, Pat Bell, Jill Carter, Denise Birkin, Pauline Dawson, Sue Johnson, Ann Buckland, Margaret Spall, Barbara Waller, Steff Bramley, Pam Gower, Lyn

Lilley, Sandra Plenty, Freda Clurow, Josie Bakewell, Linda Lingard, Lynne Marlow, Esther Johnson, Karen Page, Lesley Danby, Jackie Wilkinson, Jane Chaplin, Wendy Dear, Kim Hickling, Glenys Hartshorn, Sue Offiler, Sharon Dobb, Lynn Cresswell, Joy Chambers, Gillian Bentley, Jean Arnold, Barbara Bicknall, Janet Askey, Rita Wakefield, Joan Savage.
Supporters: Ipswich, Harlow, Welwyn Garden City, Stevenage, South East Hertford, Long Eaton, Saffron Walden, Forest Fields, Huntingdon, Derby, Bulwell East and Newham Communist Parties, Hoddesdon, Basford and Nottingham Central Labour Parties, Workers from Rolls Royce, Central Television, The Bank of England, Brixton Advice Centre and STC, Nottingham University, Sogat '82, Billericay, ASTMS, Polly Griffiths, POEU, Harry and Kay Bassnett, Dowty, Nick Franklin, Pet Coleman, Ollerton Welfare Rights Centre.

ILKESTON WOMEN'S SUPPORT GROUP

> We started in May, 1984, for the miners out from Moor Green, Pye Hill, Shirebrook, Cotgrave and Annesley who lived here. All the pits round about were closed in the 1960's and all the men moved to North Derbyshire or Nottinghamshire. Ann Jeffries and Chris Gilbert put an article in the free paper, asking for women with their men on strike to get in touch. We had some abusive calls, but there were three genuine ones.

These were from Glenys, Gloria and Maria Pringle. The three women joined the other two and started to work with Dan Rafferty and other strikers. One hundred and fifty people attended the first meeting they called. At first they were naive and disorganised but experience and increasing need were good teachers and soon the group became cohesive and effective. A separate women's committee was formed with Chris Gilbert as chairperson and treasurer.

> Food was our first priority. We asked all the local churches, you'd think they would help. We got nothing. They wouldn't let us have any of their halls. Then we thought that anyway it would be impossible for people to come to a kitchen. They live out of the centre of Ilkeston and they wouldn't be able to afford transport.

GOOD LUCK. OUR HEARTS ARE WITH YOU IN YOUR STRUGGLE

IF YOU HAVE A YOUNG MINERS BOY WE WOULD BE HAPPY TO HAVE HIM FOR A WEEK COLLECT AND RETURN.

West London. Nov 1984

To: Women's Support Group

Enclosed please find a cheque for £260 (I did a sponsored parachute jump!)

Good luck, and very best wishes

Clapton, London E.5 Nov 1984

I read youre letter in the minor which bought in Huddersfield and felt I owed at least my written support to people such as yourselves.

Every Friday night I go pub collecting for the Salvation army. Many people call us Scargill Lovers for transferring money from Kirklees to the minors.

I am proud of what we did and it's unfortunate we could not do more. As a Christian I find it impossible and only logic to support the Right to Work.

People fail to realize that Jesus of Nazerath taught socialism through. When shall people learn that only a Socialist World will be a fair World?

Many people at College and Church reckon I am to far left but Justice is Justice at the end of the day.

Thankyou Nottinghamshire striking minors for fighting my war against Greed against injustice and against a Government set to use scum tactics to break the Working man.

One day this whole Rotten World of hatred will come to an end. On that day men will question if the riches they sought to acquire by unfair means was really worth it?

Must close

Keep fighting

Keep faithful

In October part of the group split off to form a separate one at Eastwood; this made organisation easier and catering more efficient. The number of families to support dropped from 100 to 46. There was no kitchen so food parcels had to make up for it.

There was a special hardship fund for special cases and for weekly retainer payments of electricity and gas. Single men received £5 when funds allowed. In order to keep in close contact with everyone, every miner had to come to the group each week for his picketing money and women were encouraged to come along too. There was thus an opportunity to chat about any problems and give help.

October was a significant time for the group's fortunes. Contact was made with the Fire Brigades Union and Ann Jeffries, Glenys Pringle and Chris Gilbert went to London to speak at their meetings. Glenys describes how they went to ten fire stations in the East End in one day:

> Every meeting was at the top of loads of stairs. We were worn out just climbing the stairs, let alone speaking. We'd have come down the poles if they'd let us. We got tired out but it was good in a way and in three days we raised £700. We were really frightened of doing it at first, we'd never done it before. But I'm sure we got better as we got more confident. I think we could put our case really well by the end.
>
> ★
>
> We made marvellous friendships with the firemen. They gave us their clothes, socks, shoes, shirts, trousers, jackets, we had them all. Our kids got helmets and jackets for souvenirs. They used to come up here to stop and we went down there.

South Wales miners brought a transit van full of food in a generous act of solidarity. Other support came from London and helped to put the finances on a sound basis.

Ann tells of the fortuitous incident, the highlight of the year, when a seam of good coal was discovered on the local railway embankment.

> Nobody stopped the men from digging. I don't know what we'd have done without this coal, we never had enough for everybody, but this did help us.

The problem of trying to keep warm, of drying clothing and keeping the house from becoming damp was shared by everyone on strike. In Glenys's home:

> We had some lovely warm quilts. We all used to get together under them on the settee, sitting close. Folk who came to our house had to do the same, it was the only way to get warm.
>
> ★
>
> When we had jumble sales we daren't say they were for the miners. We had to find some names, make them up, so people wouldn't know what it was really for or they wouldn't have come. We had some good jumble sales.

At Christmas extra turkeys arrived from workers at Rolls Royce in Derby, from local Labour Party members and from Walthamstow Fire Station.

> We had a wonderful time, we sang our new carol, "The Red Flag". We've learned all about politics, we know what the police are for.

Ann's husband was arrested en route from his home to Yorkshire:

> He was travelling with some London firemen, through the night. The miners got arrested when they got to Rotherham. Somebody made false charges against them. I enquired about where they had taken him and they told me all sorts of stories, to worry and frighten me. I can't find the words to tell what I think of them. The children don't trust them either. Jason got a sticker from one at school, they were doing road safety. He ripped it off when the policeman stuck it on, he wouldn't have it. One little lad was having some new shoes. They were measuring his feet and the shop assistant said they were growing so fast he'd be a policeman when he grew up. He said he'd not be a policeman, he'd be a picket. He said he'd throw stones at the policemen when they beat him up.

Many women feel that their children were neglected during the strike:

> The little ones; their routine was different and they didn't understand. We'll tell them when they're older, explain why it was. We'll tell them about people like Stan and Millie who got their pensions every week and went to the supermarket to get food for the strikers and bring it to our house. They always did that before they did their own shopping and they gave us a tin of salmon or something for the raffle every week.

Now they want to repay help where they can and have already joined strikers at the Scarsdale Hospital in Chesterfield. The group is still together collecting for sacked and imprisoned miners.

Women Involved in the Group: Ann Jeffries, Chris Gilbert, Glenys Pringle, Gloria Pringle, Maria Pringle.

KESWICK ACTION GROUP

This area is in no sense of the word a community. It is really a large NCB Estate, known locally, for obvious reasons, as 'cement city'. Most of the miners who live on the estate now travel to work at the Manton and Shireoaks collieries.

From April 1984, families on strike received food parcels every week through the organisation run by Terry Healey. The group of women followed the same pattern as Carlton and Langold, holding local fund-raising events to feed children during the school holidays. Extra money came from Nottinghamshire County Council and from the Miners Union. After the October holidays the group was disbanded.

There was a great sense of isolation for the strikers and their families in this particular area of Nottinghamshire. Marlene Day speaks of the effects on the children particularly.

> If they were old enough they could understand, even if it didn't make it very much easier. There were no more treats, things they were used to. But it was especially hard for the little ones who couldn't understand. We got so sick of tinned beans and spaghetti. I'm a barmaid, I work in the local pub. I had to watch what I said

> all the time, most of the customers were working miners.
>
> It was terrible when nearly all the miners off the estate were back. I looked out of my window one day, and saw my husband at the bus-stop with four other strikers, waiting for the 'battle-bus' as they called it. They were going to shout to the men going in to work. The scabs were guarded by loads of police but they told my husband and the others that if they intimidated the working miners they'd be arrested. How could they intimidate them? They were just five strikers and about 900 men had gone back. They told them to move on or they'd be arrested for besetting. They were just standing there, doing nothing wrong. The police were terrible.

Marlene speaks sadly of the strain the strike put on personal relationships:

> If there was love, you could survive. The strike really tested that. There's a lot of trouble now between couples, those who did not both support the strike. There's resentment as well, men went back because their wives nagged them to go back. A lot feel guilty, for not standing with their mates in the strike. If you stood for principles, stood together, you've got something to be proud of.

Women Involved in the Group: Marlene Day, Ivy Spall, June Cook, Muriel Moody, Elaine King, Aileen Reynolds, Helen Hutchinson, Louise Evans, Edie Clements, Liz Morrison, Janet Sergeson.

LANGOLD WOMEN'S ACTION GROUP

As at Carlton, the group at Langold was organised specially to complement the work of the men who were already dealing with the provision of food parcels in the area. In particular, it was set up to ensure that there were funds for children's meals during the school holidays. Strikers living in the village were from a wide area of pits:- Manton, Shireoaks, Maltby, Edlington, Harworth, Bevercotes, Silverwood, Dinnington, Thoresby, Manvers, Whitwell and Dearne Valley near Barnsley. At the beginning the village had 206 striking miners;

by the end this had fallen to 79. Peggy Moulds says,

> We raised money locally with jumble sales and raffles, some people gave us donations, some workers at the *News of the World* gave £500. That was wonderful, wasn't it? Nottinghamshire County Council allowed us 30 pence each for children at school in the Whitsun holidays. Our village was just like one big family so it was easy to get helpers. June Wragge, the school meals supervisor, and Jackie Craven, her deputy, worked hard for us in the school kitchen and Pickerskills sold us food at cost. They gave us extras, they were very good.

About 80 children had meals in the school and all the time the small group was raising money and adding it to the parcel fund. Meals were provided for the children again in the summer holidays and for the rest of the strike when the school was normally closed. The five weeks of the summer were a specially good time for the strikers, the women and children.

> We had a real community there. Pat Craven, a local coal merchant, organised beer for the men at 30p per pint, sold at Smokey's Club, and this helped to keep their spirits up when they'd nothing in their pockets.

A special womens' group was organised for Christmas. Peggy, Carole Bailey, Rose Isherwood, Sharon Turner, Eva Lillie, Yvonne Price and Sandra Mould went out on a door-to-door collection. In addition to money raised from the inevitable jumble sales and raffles, money came from the Miners' Unions and from Cotham Power Station workers. All this helped the children to have a lovely time. Despite all that they accomplished, the Langold women have many regrets.

> We all wish we'd been able to get more involved. We wish we could have done things that other women did in the strike. We couldn't go out picketing, there's no local pit and we'd no transport to go to other pits. We couldn't get to the meetings of the umbrella group, that was a shame. We missed such a lot. But we still learned a lot, we were politicised. We don't trust the police any more. They

weren't impartial, they weren't keeping the peace. We saw things in this village that made us militant like we'd never been before. At Christmas the police were driving around in a van with a Christmas tree and all streamers and decorations telling the children they wouldn't be having much of a Christmas. Imagine telling them that, it made us aggressive, treating our children like that.

One of the greatest regrets is that their village was divided. They had previously valued their close-knit community and it hurt them seeing it torn apart in the strike.

Women Involved in the Group: Christine Wright, Susan Parkin, Jean Ramplin, Margaret Goulding, Susan Nash, Peggy Moulds, Carole Bailey, Rose Isherwood, Sharon Turner, Eva Lillie, June Wragge, Jackie Craven, Yvonne Price, Sandra Mould.
Food Parcels: Terry Healey and Bill Brown.
Supporters: Harold Plant, Joe Ashton, Langold Labour Party, John Tibkler, Mr and Mrs Cooper, The Hilltop Club Committee, NUPE domestic workers at Kelton Hospital, old people in the village, News of the World EEPTU workers, Rosie Walker and Ruth Griffiths.

MANSFIELD (LADYBROOK) SUPPORT GROUP

People working here had the greatest difficulties of all the groups. The group was not based on one pit, but supported miners' families living in the town. When at work the men were employed in the Nottinghamshire, North and South Derbyshire and South Yorkshire coalfields. The District Council would not help with obtaining premises and apart from the Nottinghamshire Area NUM no other labour movement organisation had offices in the town.

In the early days food parcels were made up in people's homes. Then the Quakers came to the rescue and gave the use of the Friends' Meeting House. All the facilities of these premises were available and this generosity made a great difference to the group's work. During the whole period there was no charge made for rent or electricity. Their support was wonderful and much appreciated.

The premises were not spacious and Friday was a particularly hectic

MANSFIELD (LADYBROOK) SUPPORT GROUP

Tuesday/Wednesday/Friday : 10.30 a.m. to 3.30 p.m.

GENERAL INFORMATION

HOLIDAYS

Please contact MAUREEN at Mansfield Centre, if you would like details of free holidays which are available.

FAMILIES with children at school, receiving free meals, can obtain clothing vouchers from the Education Department. (£5.00 per child).

All contact made through our Centre is private and confidential.

FORTHCOMING EVENTS

1. 'Sponsored Walk' - from Mansfield Centre!! VOLUNTEER WALKERS NEEDED! SUNDAY, 29th July, 1984.

2. FREE CAMPING HOLIDAYS - throughout July and August! Available from Mansfield Centre; let us know if you wish to take advantage of this generous offer.

APPEALS FOR HELP

1. We have a **knitting machine** at our Centre, but we don't know how it works! **Can anyone help, please?**

2. Is anyone able to do some 'baking at home' for us? Mince pies, jam tarts, etc. (small amounts, ingredients supplied, when available).

3. Materials, and sewing machines, are waiting to be used at our Centre. We could make clothing, soft toys, etc. Can anyone spare a couple of hours?

4. Any '**FUND-RAISING**' ideas would be welcome.

5. We have requests for assistance with 'car maintenance'.

   Do we have any typists, amateur photographers, cooks, child-minders and bottle washers, who would be willing to lend a hand?

6. We are hoping to have a 'CHILDREN'S PARTY' during the school holidays, but we need to raise the funds to do so. PLEASE HELP.

THIS IS YOUR CENTRE! MAKE USE OF IT!

GET TOGETHER: STICK TOGETHER: WE WILL WIN TOGETHER!

day. At one stage, 350 parcels were being collected. People met, talked, joked together, recounted stories of the strike and enjoyed food, especially the delicious home-made cakes etc. provided by Mr and Mrs Wood.

Only a small handful of people adopted this group and so it was heavily dependent on money allocated from central group funds. Maureen Gill, the chairperson, repaid the central group by speaking at meetings and other events and raising thousands of pounds.

Women Involved in the Group: Maureen Gill, Fay Calladine, Jackie Naylor, Annette Needham, Dawn Betts, Bridie Topliss, Sharon, Doreen.

MANTON WOMEN'S ACTION GROUP

Manton is a Yorkshire Area pit, situated in the county of Nottinghamshire. At the beginning it had 800 strikers, of whom 187 were left at the end.

The impetus for the group of women who worked here was the need to provide meals for children during the Easter holidays at the beginning of the dispute. Nottinghamshire County Council gave the use of Manton Primary School as a canteen for the children of striking miners. Only the premises were available, no cooking facilities were provided so the women cooked food in their own homes and it was taken to the school. Barbara Scott collected the food in her car each day, enough for 200 children.

> Then we wanted to go on, with a permanent kitchen, so we got the Manton Miners' Welfare Club committee to give us the cricket pavilion. The kitchen was tiny so at first we did snacks at mid-day. Then we borrowed two small electric cookers and a camping cooker from the colliery. People brought their pots and pans from home and somehow we managed. We could do a lot more then, though it was really cramped.

When Manton was mentioned in *The Morning Star* cheques and money orders from individuals poured in and formed the basis for funding for the duration of the strike. Nottinghamshire County Council gave money for children's meals and Manton Colliery hardship fund donated £80. Some local shopkeepers reduced their prices. Jimmy Munson and Mr Patel were particularly kind; often they gave tins which had lost their labels and the women had fun guessing what was inside. Occasionally it was dog or cat food—still welcome. A butcher sold his meat at low prices and some local farmers gave sacks of potatoes.

> About two dozen women were involved altogether though Glenda and I were in the kitchen on our own for most of the time. Then in the last two months Janet Bowering and Ann Cooper came and helped. The men peeled potatoes, especially when they had to keep away from the picket lines. We were doing 70-150 meals every day. Pete and Norma Taylor did the food parcels. We were too busy to go picketing though Ann Cooper went regularly to Manton pit.

Regular meetings were held in the Miners' Welfare when problems and grumbles were discussed, together with plans for the coming week. Joint meetings were held with women from the neighbouring Shireoaks Women's Action Group.

> There wasn't a formal arrangement of the group, but we had a delegate, Norma Taylor, to the central group and we had donations from them whenever we were short.

Women Involved in the Group: Barbara Scott, Glenys Hawkins, Janet Bowering, Joan Bagshaw, Ann Cooper, Eileen Watkinson, Valerie Singleton, G. Chambers, B. Gaskill, Karen Street, Veronica Hayes, Susan Davies, C. Greasley, J. Hurt, J. Turner, Maureen Baxendale, J. Potts, Maureen Tindle, L. Percell, S. Westley, Rose Ogley, N. Billham, J. Hallam, Sharon Jervis, S. Jervis, Vera Whitehouse, Mrs. O'Sullivan, Eunice Goodhall.
Supporters: Walthamstow, Chipping Barnet, Ealing, Grove Green and Leyton Labour Parties, Stratford Polytechnic, Philip Headley at the Stratford Theatre Royal and the Southall Sikh Community.

MOORGREEN SOLIDARITY GROUP

We broke away from the Ilkeston group because we thought we would do better if we had two smaller groups. We had 61 miners when we started, then 54 at the end.

Weekly parcels were packed in Dorothy Naylor's garage in Eastwood Avenue. She describes fitting it up with shelves for the storage of food and for second hand shoes.

Food came from Corby in carrier bags and Whalley Bridge Support Group in Derbyshire also sent food rather than money every week. I think supporters sent food so that they knew their money was really being used that way. But it was a lot harder, trying to sort out parcels that way than being able to go to the supermarket and buy, say, fifty tins of soup. It was really difficult sharing things out fairly.

Sheila Davies, the milk woman and Maggie Green, the postwoman, did their early morning work and then came to help Dorothy pack the parcels. A gardener from Moorgreen Colliery who was due for retirement, so had to continue working, gave £50 from his wages every week. Ted Moore and his friends gave about a hundred items of food every week and there was a system of saving 'specials' until there was enough to give everyone something like coffee, ham or tinned fish, extra to the basics they had every week. It took weeks to collect these foodstuffs but so important were they that an announcement of a 'special' was always made at the weekly meetings. On parcel lists names were recorded with details of numbers of children or any other category in special need and there were extras like rice pudding when these were available. The women in the group got to know the needs of the people on strike and were often able to send out good clothing, sorted before a jumble sale, to particular families.

We had no kitchen and no hall for regular meetings. We used to meet down on the picket line at the Gotkin open-cast site. We put easy chairs in the picket hut and we had carpets on the walls as well as the floors. We used to take onions and burgers and sit drinking Ovaltine and talking till the early hours. That was where we

> talked about the strike and the political situation. It was a very special place to us.

Members of the group worked very hard and are proud of their efficiency and their success. Women spoke at meetings and collected in the street.

> I went to Ipswich, collecting with supporters. Can you imagine how I felt when they gave me a placard to wear round my neck? I was so embarrassed but I daren't show it because they weren't a bit bothered and they were doing it for us. That was the hardest thing I had to do in the strike.

Dorothy feels that there were more good times than bad and recalls especially the socials in people's homes with home-made wine and beer and whatever could be found to make a party. Christmas was a wonderful time, especially for the children. People came from Corby and miners and their families from South Derbyshire joined the group for a party at Langley Mill Cooperative Hall. On December 21st another party with a superb buffet, raffle prizes and drinks made everyone believe they had a better time than the working miners. There were so many children's presents over that six boxes were sent to Dr. Barnardo's Home.

Amongst the helpers were two women who came regularly at 4.00 am to picket at Woodlinkin open-cast site. They brought trays of food and steadfastly refused to give their names. They were known as 'the ladies of the night' and no-one ever found out who they were or where they came from.

The most unusual fund-raising event was a dog show, organised on the Shipley Boat pub field. Local residents asked for a repeat show the following year. Ann Fownes and Addie Wilkinson became known as the 'jumble sale queens':

> The monthly sales at Kirkby-in-Ashfield were good and sometimes we got a stall on Heanor Market. We had to give all sorts of names of groups to get the stall, we wouldn't have got it if we'd said it was for the miners. We had everything on the stall, anything we could get hold of to sell; we've stripped our houses, there's nothing left to sell.

The advice and money which were always available from the central group was very important to the group. Women have made many friends and have gained in confidence. Despite the times of anxiety and sadness, they say, "We wouldn't have missed the year for anything".

Women Involved in the Group: Dorothy Naylor, Addie Wilkinson, Ann Fownes, Wendy Skelton, Maggie Cryer, Margaret Rafferty, Sheila Brown, Cath Fowkes, Sheila Davies, Pauline Maxwell, Kath Moseley, Carol Spendlove, Sherry Clay, Gloria Mee, Hazel Roland.
Supporters: Newsagent in Kirk Hallam, gardener from Moorgreen Colliery, local trader John Wright, Fred and Nancy from the Erewash Hotel, Maureen Moss, Pat Skemp, Forest Fields Support Group, Bradford College, Trevor Sarginson, local Labour Parties, Madeline Whitam from Nuttall, South Manchester DHSS offices, London teacher Bruce Goddard, NUR member Brian O'Neill, Ian McCormack from the local Ecology Party, Pam Elliot from Berry Hill, Mr and Mrs Sutton from Long Eaton, Lesley Bracey who donated marmalade and onions, Mr. Wild who sent jars of home made jam, Joe and friends.

NEWARK MINERS' SUPPORT GROUP

Before we actually formed our group on 23rd May 1984, we had voted a donation of £100 from our local Labour Party branch to the Nottinghamshire Miners' Forum and members were already collecting money and food for Sue Taylor to take to Ollerton Support Group.

A small group of activists, mainly officers of the party, decided that there should be a planned programme of fund-raising and that we should put the group on a permanent basis. I went with Jef, my husband, to Blidworth to hear what they were doing and where our help was most needed. We were advised to adopt Bilsthorpe and Rainworth and so it was that our firm association with these two groups was established. From then on, Jef and I went about twice a week, with a car load of food, clothes, shoes, and anything else useful, as well as donations amounting to about £25 to each group.

We needed to find out if there were men from Newark on strike.

This proved to be a very difficult and sensitive task, but once we had found one, we began to hear of others. The 200 or so miners living in the town are employed at pits scattered all over the northern coalfield and no-one seemed able to provide us with names and addresses. We were afraid that if the men were not located quickly, hardship would drive them back to work. We eventually found fifteen, ten of whom joined the group as the others went back to work. We lost a further three during the summer but seven remained throughout and became known as 'the magnificent seven'.

Having formed our group and found the strikers, it was necessary to raise money. Attempts to obtain permits for street collections failed. Strikers collecting outside local factory gates in Newark soon learned that support in the town was minimal. Newark Town Labour Party members were urged to pay the levy of 50 pence per week. Some did, often paying much over the levy; others gave at their work places. We began to organise the usual fund-raising activities—dances, concerts, raffles, barbeques, jumble sales and car rallies, but we could not advertise them as being in aid of the miners because of hostility in the town. We placed a paid advert in the local paper asking for help and though response was very small, we made contact with a few people who gave wonderful support for the whole year.

In September, seeing the cold winter ahead and the possibility of our funds running low, I wrote letters to *Labour Weekly,* the *Morning Star,* and the *New Socialist,* on behalf of the central group, appealing for knitting wool and knitted garments for children. The response was overwhelming. The postman was laden with sacks every morning and the hall of my house was packed. As fast as we distributed the wool and the clothes, more came in. Wool came from as far away as Sweden. Mrs Berry from East Finchley began to knit lovely white baby vests and sent them regularly throughout the strike. People sent scarves and mitts for the pickets as well as beautiful clothing for the children. Others, unable to send wool, sent cheques. We had offers of holidays, all expenses paid. David Moulson's mother began to knit and went on for more than a year until all the wool had gone.

We joined in with the central group in the early weeks when we needed funds but then we became self-sufficient and paid back sums of money when we could. Our delegates attended the meetings and reported back to our own group meetings held every Thursday evening at the home of Zena and Kevin Sanigar. Some of our women joined in the women's pickets. Edna Colston was shaking in her shoes at her first picket at Thoresby:

> I had my period and I stuffed my handbag full of Tampax just in case I got arrested; I found that Judy Sim had done the same. We seemed to talk about nothing but the strike at our house and every morning my little girl, Caroline, would ask if we were going marching. I've got four children. They all knew about the strike, we were working for it all the time. I was worried sick that I'd get arrested if I went picketing.

I went to London to speak for the central group, to the GLC and Hendon Labour Party. I don't think people really wanted to hear me though, they wanted a miner's wife. Even though I could speak with a lot of knowledge and experience, I was not the real thing. However, when Jef and I went for a week to stay with a relative in the Isle of Wight, we spoke to members of five Labour Party branches on the island and came home with our car loaded with food and clothes. I spent a good deal of each week going round second-hand clothes shops in Newark buying children's clothing, usually with a list of names and sizes for individual children in need.

Although the main purpose of our group was to raise money, which we did very successfully, another important goal was to give personal support to the Newark miners and their families, living so far away from their pits, isolated in a generally hostile town. We came to know them well. We had hilarious nights together at the Vine, especially playing charades. We felt great respect for them and our friendships grew as we worked with them for the period of the strike. We experienced socialism in action.

Women of the Group: Edna Colston, Andrea Eason, Jill McCauley, Val Picker, Mary Roberts, Gail Helliwell, Zena Sanigar, Joan Witham.
Men of the Group: Richard Colston, Dave Barton, Merf Roberts, Jack Probets, Les Smith, Jef Witham, Kevin Sanigar, and
Striking Miners: John Eason, Frank McCauley, Alistair Welsh, Bob Morman, Chris Martin, Dave Moulson, Gary Fearon.

Supporters: Mrs Redfern, Renee Le Petit, Portland Labour Party, Joan Bourne, Mr and Mrs E. Nice, Margaret Moulton, Valerie Marker, Iris Handscomb and Coulsden Labour Party, Bill and Kit Fisher, Kit and Jack Probets, Stuart Scrimshaw who collected from his workplace every week, Newark Town Labour Party members Mary and Fred Turner, Olga Newstead and family.

OLLERTON WOMEN'S ACTION SUPPORT GROUP

This was one of the pioneering groups in the county and its early experiences gave significant guidance to others which were set up later in the year. Ollerton was the location of the Miners' Forum Strike Centre and of the legal and welfare rights organisations set up for the strikers. It was therefore a focal point for all strike activities.

Ann Nesbitt, Rhona Chapman and Linda King set up the group to cater not only for local miners, women and children associated with the pits at Ollerton, Bevercotes, Clipstone, Thoresby and Bilsthorpe, but also for the many pickets who came from Yorkshire, Northumberland and Scotland to try to bring out the Nottinghamshire working miners.

Every Tuesday night an open meeting was held and anyone needing advice or information on making contacts for financial support was welcome. Ollerton had made contact with Norwich Trades Council and Labour Parties early in the strike and, reasonably secure, were able to offer help to other groups less well estabished. When women were asked by Ida Hackett from the central group to share their contacts and expertise, they unanimously agreed to send speakers out from Ollerton to raise money for the central group funds. Brenda Greenwood and Margaret Hume travelled to meetings all over the country doing this work. ''But'', they said, ''our links with our own supporters were so firmly cemented, we weren't going to give them up, and the people from Norwich wanted to stick with us as well.''

> At the beginning we were producing over a hundred meals a day, with breakfast and a mid-day meal five days a week. We had a kitchen in the Ollerton Miners' Welfare and we had a good cellar where we could store our food and pack parcels. We didn't have

> to pay rent but we had to pay for electricity. We borrowed catering equipment from John Hand, a local butcher, and somebody lent a refrigerator. We all brought our own pots and pans, we weren't doing much cooking at home anyway.

The Centre became a second home for the strikers, women and children. Members of the Hardship Committee were usually on the premises, so a large number of people constantly visited Ollerton; it was probably one of the busiest centres in the county.

> At first we met together for about half an hour every morning to plan the day's work, but when it was going well we dropped this off. Ann Nesbitt was in charge of catering and worked in the kitchen every day with about twelve others helping. But as time went on it became too heavy for women to keep coming every day. They had no time for their own work so they came on different days and had some time at home. Kath Eddy took charge of food parcels.

Women like Kath Black, an NUM member on strike from Ollerton pit canteen, went on the picket line at four o'clock in the morning before coming to the kitchen to start work. Ollerton pit was the scene of heavy picketing in the early months of the strike. It was there that a young miner, David Jones from Yorkshire, was killed. Much money and food flooded into the village after this event as people responded to the scenes of strife and anguish shown by the media. It was at Ollerton that Brenda Greenwood was arrested on a picket line and sent to Risley Remand Centre. She comments:

> We have learned many political lessons during this year, we know now what governments will do to break industrial action. We've become very militant and radical from what we have seen.

Initially some of the men were chauvinistic towards the women, possibly due to jealousy, but as they saw their efficiency and success they were forced to admire their drive and capabilites, grudging though this admiration was at first. Relations were amicable though each group was independent until about Christmas, when the men had to ask the women for financial assistance and it was decided to join forces.

Brenda Prince/Format

Christmas Party at Ollerton, 1984

LADIES

A FURTHER SMALL CONTRIBUTION

BUT NOT TOWARDS A PRESENT
FOR THE NEW ROYAL BABY
!!!

WHATEVER HAPPENS YOU
HAVE ARRIVED . SO STAY
IN THERE

GOD BLESS

Despite their overwhelming gratitude to those who helped them during the strike the women say that they really needed more than money and food; they also needed the fullest backing from trade unions which meant all workers coming out in support.

Women Involved in the Group:

Chairperson:	Brenda Greenwood, then Margaret Hume
Vice Chair:	Denise Appleton
Secretary:	Sandra McGinty, then Brenda Greenwood, then Linda King, then Esther O'Regan
Treasurers:	Rhona Chapman, then Ann Jackson
Trustee:	Janette Tomlinson

Other Women: Kath Black, Ellen Steel, Viv Johnson, Kath Eddy, Ann Nesbitt, Diane Baird, Katy Lees, Moir Green, Mavis Dixon.

RAINWORTH FAMILY SUPPORT GROUP

When women came together to form this group, they tried to get the local Drill Hall for their premises. The Rufford Pit manager refused permission. The Youth Club pavilion was offered to them but it was totally inadequate and the women refused. They decided to move into the Drill Hall without permission. It was decided to stage a 'sit-in' and men, women and children from Blidworth, whose own 'sit-in' had proved so successful, came to help in the occupation. The squatters stayed all day and the next night. Even though the police came at midnight they refused to move. They were asked to leave again by members of the management committee, but stayed firmly in possession. Television cameras and reporters gave them unwelcome publicity but the occupation was a success and, without ever getting permission, they kept part of the premises for the duration of the strike.

There was a tiny kitchen and dining room upstairs and a meeting room on the ground floor. All provisions had to be carried upstairs, as well as prams and carry cots so that women had their children nearby when working in the kitchen or stores.

Facilities were far from satisfactory but the women managed at first with two Burco boilers, one for potatoes and the other for stew. Later they acquired a microwave and a full-sized oven. Numerous regulations were made by the management but were largely ignored. The place became home and gradually people who were strangers before the strike grew to know and work with each other. Inexperienced in working collectively, some women found it hard to work under another woman's leadership and direction. After a number of disagreements about how things should be done, some left and only a small group of women remained to do all the work. Then a striker, Ernie Daglish, offered his expertise and together they worked harmoniously.

Only one woman, Annette Needham, felt capable of going out speaking to raise funds. When she moved to the nearer Mansfield group there was no-one to go. Twinning arrangements made with Lenton, Nottinghamshire, and with Cambridge, brought in about £130 and Newark sent all the help it could. Newark also raised a great deal of money to buy clothes for children and underwear for strikers and women. There were new jeans and trainers for men who needed them.

Although the women inevitably learned political lessons as a direct result of their experiences at meetings, marches and rallies, they were too busy running the kitchen, doing part-time jobs or caring for children, to become more deeply involved. For the number of women actively involved, the group was one of the smallest in Nottinghamshire.

Lynn Telford was one of the women who stayed active throughout. She describes a terrifying experience:

> We were just settling down to watch a late movie on TV. I was near the end of my pregnancy and my little boy, who's five, was in bed asleep. Richard had been out to the club and he'd had an argument and a bit of a fight about the strike. He had tried to tell the other man about loyalty and solidarity but he didn't want to know. Richard had only his trousers on when suddenly this man forced the door, came in with a bow saw and a great big knife and started slashing at Richard, breaking the furniture and slashing the settee and chairs. I was terrified, he was slashing at his arms, his neck and all over his back, there was blood everywhere. If this had been the other way round, Richard would have been sent to prison and he'd have lost his job. The man who attacked him got off lightly, he didn't get sent to prison.

Women Involved in the Group:

Chairperson:	Annette Needham, then Jan Pearce
Secretary:	Claire Godfrey
Treasurers:	Chris Alsop and Joan Spencer, then Karen Coupe and Ernie Daglish

Other Women: Morell Inskip, Liz Gordon, Lynn Telford, Christime Kulek, Linda Kulek, Elaine Overton, Nell and Karen Marsh.

Supporters: Cambridge Miners' Support Group, Newark Miners' Support Group, Mary and Fred Turner, Radford Raft Group, Herne Hill and Willesden Labour Parties, London, and Mr. Fuest Wang of London.

RETFORD MINERS' RELIEF FUND

Our miners in Retford were on strike from pits at Bevercotes, Manton and Armthorpe. They were living in Retford, Tuxford, Elkesley and Worksop. Jeff Corrie asked the officers of Retford Labour Party if we could use their small hall and we got it for a kitchen.

Janet Bland remembers how she was told by Jeff that if she went along to the kitchen she would find lots of helpers there.

He was just saying that to get me to go. He only knew about two others, Jane Elkington and Pauline Stoddard. At that stage there were no cooking pots and pans, no cutlery and crockery, no cooker. There's a second-hand shop next door to the hall so we bought what we could with £30 from the Hardship Fund. People lent what they could. I'll never forget the first meal we cooked, it was for about forty people, soup or stew made with bones from the butcher nearby and vegetables thrown in. We cooked this concoction in an old copper boiler and it lasted three days. We hadn't much money so we weren't doing parcels then.

Trying to collect in the town was a fruitless exercise, though workers at Babcock and Jenkins gave £15 to £20 every fortnight and local bakers Couplands and Bamfords donated buns and bread daily throughout the strike.

One of our members did some voluntary work for which she was given a small ~~honorarium~~ honorarium, & she has passed this on to send to you. She asked particularly that, if possible, she would like you to spend this money on some good shoes for those children most in need. I have great pleasure, therefore, in enclosing a cheque for £75. With it comes the best wishes of our members who hope you will soon achieve the victory for which you have all fought so hard & so long, & which we know is for us all.

Ilford Pensioners' Action Association. Nov 1984

As an old miner still embittered by memouries of the ~~20~~ twenties and thirties I am pleased ~~so~~ to ~~a~~ send a little to help the miners in their struggle to survive. together with trade unionism as whole which is under attack from this, Tory Govt. I enclose cheque for £5.

Consett, Co. Durham. May 1984

> Then a group of Welsh miners came and saw how badly off we were. They raised £300 for us, so we could have better meals. More and more people were coming to the kitchen, though some never came at all. We got a cooker and two hot plates on a calor gas stove and more money was coming in. Men were also going back to work so we were better off and began to give food parcels.

In June miners came from Scotland to picket at Bevercotes. When they returned to Scotland they took four of the Nottinghamshire strikers with them, returning a week later with a transit van full of food and a donation of £1,000. These miners decided to adopt Retford and thereafter when it was needed, money came from Fishcross, the main strike centre in Scotland. This generosity eased worries about finance, though there was never really enough to go round, especially when winter came and people needed fuel. Bassettlaw District Council gave an allocation of money for food, £60 for eight weeks, then a further £2,000 near the end of the strike. A local co-op allowed a 5% discount on food purchases and some donations came from pensioners. A trainee manager at a local pit secretly brought his wages to the group every week and must have donated about £2,000, with an additional sum of £200 specially for the children to go for a day to the sea-side and for a party at Christmas. His two brothers, also trainee managers, who were threatened with dismissal unless they returned to work, also sent donations from their wages every week.

There was a formal committee of four men and four women and though relations were usually good, there was some disagreement over food parcels and their contents. The men wanted to have a qualification for receiving parcels—that the men should have to go picketing. The women disagreed and demanded that everyone should have a parcel. As the number of men on strike decreased, so the number of women helpers was also reduced. The work began to take a toll on the women left. They had their own homes to clean and their children to care for. There were endless jumble sales, discos and raffles to organise and though the men were supportive they were heavily involved in picketing. In addition, some were working on extensions to the Retford Labour Hall.

Janet regrets that they were not able to be as involved in all aspects of the strike as some other women were:

> We wanted to go picketing but it was just about impossible, and some of the men were against it anyhow. The best times were in Scotland with other folks on strike, and on marches and rallies. The worst times were when the men were going back, even though they said they were solid for the strike. We were worried that we'd lose the Labour Hall. People were being threatened with having gas and electricity cut off. We'd had men arrested and imprisoned. Things looked really bad.

The worst time for Julie Corrie was when she came out of hospital with her first child:

> I knew my husband was heavily involved in the strike, but he'd been arrested five times on picket lines. They took him to Lincoln jail, then they released him. I knew he'd be arrested again if he went back to a picket line and I wanted him with me and the baby. And he wanted to do both. I really wanted his help and I was frightened.

Seeing the problem of a man torn between loyalty to his union and to his wife and child, the women rallied round and someone came to see Julie every day, giving her practical help and care and talking over her personal problems and anxieties. Julie recovered from the birth, supported by the other women, in peace and security, whilst her husband was able to carry on with his strike activities.

The women who were left together at the end of the strike are now firm friends.

Women Involved in the Group: Janet Bland, Deborah Jackson, Julie Corrie, Anita Durham, Pauline Stoddard, Pauline Alport, Margaret Sinclair (who did a great deal of cooking at home for the group), Jane Brett, Janet Sherman, Cathy Tisdale, Michelle Tisdale, Diane Mannerfield, Teresa Henshaw, Jane Elkington, Terri Lacey, Denise Oglesby.
Supporters: Pensioners, local trainee pit managers, Retford Labour Party, Chelmsford women's organisations, Child Poverty Action Group in London.

RHODESIA WOMEN'S ACTION GROUP

In April two women began providing parcels from money raised through jumble sales and raffles at the Miners' Welfare at Rhodesia. There were also plans to use money for discos and street parties until it was realised that it was urgently needed for food and that this had to be the priority. Terry Healer, a striking miner, brought about twenty women together to thrash out the problem and it was decided to go out and raise money rather than trying to rely on local contributions. At a meeting in Ollerton the Rhodesia women met a councillor, Sue Carlisle, from Stepney Labour Party, and she promised to make some contacts for them in London. The result was a link-up with Bow and Poplar Labour Parties, especially with the women there.

Ann Burrell and Vicki Smails went to speak at meetings and to collect on the streets of London's East End.

> We were very worried about what kind of reception we'd get. We expected people would be too poor themselves to give us money, but they were grand. They hadn't got much for themselves, they were underprivileged and exploited, but they were warm and friendly and so worried about us and our families. It was a marvellous experience to meet them.

Ann explains that extra money from London meant better parcels.

> We had women at home working to make money and we were out all over the place doing the same. We did so well that by July we could have a kitchen. We got the Bowls Pavilion at the Rhodesia Miners' Welfare. Everybody was good to us but cooking was a real problem with only a single cooker to do a meal every day for 60 to 100 pickets.

The local school was used for meals for children during the holidays. This meant that there were two kitchens to run, and most of the women had to stop picketing. Men continued to picket rather than go into the kitchens to help out because their picket pay was an essential part of the family income.

More and more people came for help as their savings ran out and it seemed that there would never be enough funds to help everyone. A special appeal was launched and men from Shireoaks and Manton marched to London to raise money for the children. Everywhere along the route supporters gave them hospitality. The first stop was in Newark amd a party was held at the Vine with Kevin Baron as guest speaker. The men stayed overnight with members of the Newark group and with other friends.

Some of the previously active women found jobs to keep their families and an even greater strain fell on those who stayed in the group. Women from Rhodesia went to Greenham Common to see for themselves the strength and commitment of the women there, and others joined in the special 'Mines not Missiles' march which came to Nottinghamshire.

> We marched in solidarity with them. We met peace women and we saw the link between their campaign and ours. We met women of the 'Lesbians Against Pit Closures' group and we heard their stories and their problems. They have a rotten time but they don't give in. Some of our men thought we'd give in when we met the first bit of bother. They didn't think we could achieve miracles but I think we did. We kept the men from having to go back. We didn't starve and we should have done.

Vicki Smails feels that the worst time was in November when there was a local initiative to get men back to work:

> We knew what was happening, it was awful, men were running from house to house, in and out they were. There wasn't anything we could do, just watch. It was all organised and though we knew what was happening, we just couldn't believe that the men would go back. It put us right down in the dumps. We felt ill. The only thing we could do was to try and think about Christmas and get ready for it. We knew Christmas didn't matter that much, for ourselves, but it did for a lot of families, because of the kids. If they could get through Christmas, into the new year, they thought they'd manage to go on. We had very little coal, but we had presents for everyone and special parcels. We had the time of our lives.

Like so many of the women, those from Rhodesia had a year of intense political education. They have been very strongly influenced by the feminists they met; they have joined the Labour Party and are taking office. Ann, who wrote poetry during the strike, hopes to take a course in Communication Skills at the local college. Vicky was offered a place at Northern College for a three year course in Social and Community Studies. When her daughter has completed her 'O' level course she will consider this offer seriously. Together the women have discovered a great ocean of untapped abilities and together they are exploring their new strengths.

Women Involved in the Group:
Chairperson: Irene Coe
Secretary: Vicki Smails
Treasurer: Denise Bradford, then Yvonne Gilfoyle
Other Women: Elaine Callahan, Ann Burrell, Jean Coper, Joan Lincoln, Linda Sherwood, Carol Hunt, Cilla Brown, Sylvia Keyhoe.

RIPLEY LABOUR PARTY SUPPORT GROUP

It was expected that Chesterfield would take the Ripley group into its organisation for support but an early appeal was not accepted, though a donation of £250 was made. And so it was that Ripley in Derbyshire was taken under the wing of the Nottinghamshire central group and received money and food when it was urgently needed.

At the beginning there were 32 men living in Ripley on strike from Blackwell Cable Works, Newstead, Annesley and Shirebrook pits. At the end there were just 4. Margaret Eley tells how a phone call to the secretary of the Ripley Labour Party, Christine Worth, telling her that a number of South Wales miners were coming to try to stop the movement of coal nearby, in Denby, was the impetus for starting the group.

> Within a week we decided to set up a group and try to find out where our local strikers were and what they needed. Money came from the levy on party members and from lotteries, raffles, etc. We had no help from the Labour-controlled Ripley District Council and relationships were soured as we tried to get their support. Our

street collections raised so little that we decided to put our energies to better use. We received terrible abuse. The most productive events were public meetings with Tony Benn, Dennis Skinner, Peter Heathfield, Ian Micardo and Joan Maynard as guest speakers. We raised a lot of money that way. When women from the Welsh coalfield came to speak, meetings were well-attended and collections were good. We had money from other Labour Party branches in the area: Heanor, Codnor and Loscoe, and occasionally from Little Eaton. Parcels were packed and distributed from my home every Saturday by a team of volunteers. The local vet across the road let us pack his garage with our stores. As a demonstration of our solidarity we used to have a rota for picketing at Markham and Arkwright pits at 6.00 in the morning.

We were disappointed that we did not get our strikers and their wives very much involved, but for us, this was the crystalisation of our socialist principles, our belief in basic trade union rights and civil liberties.

Women Involved in the Group: Christine Worth, Margaret Eley, Janis Gregory, Jo Allen, Heather Benbow, Lorraine Mullings, Margaret Kent, Moira Wales, Diane Renshaw.
Men Involved: Clive Eley, John Bailey, Mike Frend, John Stratford.

SHERWOOD WOMEN'S SUPPORT GROUP

Tina Warboys started the group and managed to acquire the Labour Hall at Mansfield Woodhouse for a centre. At first soup, tea and toast were provided for miners and their families on strike from Sherwood, Crown Farm and Welbeck pits, and the hall was open from about 5.00 am until 2.00 pm.

There was a great deal of opposition to the use of the premises by some Labour Party members and there were constant niggling complaints about dirty marks on chairs and on the radiators.

Although other people used the hall and the kitchen, we always got the blame for damage and broken pots. It was always us, though

> we cleaned and scrubbed everything every day. We knew they were trying to get us out so we were specially careful. They wouldn't let us have the place on Mondays. They said it was wanted for other, more important things—like a carpet sale! There was always some reason for not letting us in on Mondays. We dreaded that one day they'd say we couldn't have it at all.

Jane emphasises how important the meeting place was. There was always a good atmosphere in the kitchen and a lot of fun.

> You had to be good to get on from gravy to custard making, it was a great honour! There's not very much fun in cooking great bowls and jugs of gravy and custard, but somehow we made it funny. We used to get in each other's way in the little kitchen but we didn't get on each other's nerves, we had a laugh instead.

No parcels were packed at Sherwood, they came from the Mansfield group. As the strike progressed, women came together and felt like a family group. They began to go to the picket lines and meet women from other areas.

> It was like a grape-vine for news. We got to know what was going on in other groups. The picket lines were often very frightening, we used to sort of tingle with nerves, hoping we wouldn't get hurt.

Maggie tells how sad the women were when faces disappeared from the kitchen:

> We knew they'd not be coming back, they'd gone back to work. There were only 37 left at the end, we watched them drift back one by one, then a lot before Christmas.
>
> When we had a birthday we made a special celebration, we needed to, it kept our spirits up, especially for the children. We had discos and parties and a marvellous benefit night at Stewarts Banqueting Suite. We did all the catering for that, I reckon we were good enough to start up our own catering business by the end!

Sorry to be a bit later than usual. I've just had a short spell in hospital (nothing serious) and today is my first day at home for over a week.

I rejoiced over the acquittal of Clive Ponting so much I nearly tore my stitches cheering. In unexpected places you find people who are sick and fed up with that dreadful woman and her arrogance. She is due for a humbling lesson and it is on the way I am sure.

Meanwhile the struggle goes on and you and your brave people are making history and maintaining a tradition of working class resistance to oppression.

History never repeats itself exactly but there are many examples in the past which show that what you and the miners have done and are doing is in line with all that is best in the history of our country.

One fine day I hope I shall be able to make a trip to Mansfield and see you to say my thank you in person. Finer weather than this will be needed! It's perishing cold here.

All the best,

> Trips to Kent and Scotland were good. We went to Polmaise, the people were marvellous, we were kind of special—from Nottinghamshire and on strike. We lost a lot of our friends from round here, school friends, women we'd known all our lives. We stopped speaking because their men were scabbing. Life-long friends didn't want to know us because we were on strike. I met one of them in a shop and she said to me, "What's a nice girl like you doing, involved in this strike?" It hurt, that did. We stopped going shopping; we hadn't any money anyway, but we didn't want to meet them.
>
> But we found new friends. Look at Connie Birkin, how good she was to us. She brought flowers for my father's grave, she knew I couldn't afford them; and she made toys for the kids and dressed old dolls in new clothes, they were lovely. We made hundreds of friends. We went collecting to Liverpool, St. Albans and Hatfield. We went speaking to London and Manchester. We got a lot of surprises, we didn't think people would help us so much, didn't think they cared about miners, especially when they'd never seen a mine. We stopped judging people by how they looked and what they'd got. You never can tell.

It was exciting for women to be interviewed by journalists from Swedish Television and to welcome visitors from all over Britain, America and France. Children from the group went for holidays to Belgium, Holland and France.

> Their names had to go in a hat if too many wanted to go. The central group did this and made all the arrangements. It was a chance in a lifetime for our kids. People really looked after them and they all had a great time. It was a bit of a worry though, letting them go off to strangers. We were always glad to hear they'd arrived safely and were all right.

The women try to explain how they have changed:

> We hadn't any experience of trade unions before, we knew a bit about the NUM but not that much. We weren't political, some of

us were even a bit on the right. We've moved to the left, we've listened to Tony Benn and Dennis Skinner and to the leaders of the NUM and we look at things differently now. It's been an eye-opener for us women. We must have been blind before the strike. Never again, they won't take us in again.

★

We've got a new kind of freedom and we're holding on to it. There are all sorts of problems now, our men are on 'mixed' shifts, we can't get transport and there's nobody to baby-sit if the men are at work. There's all the debts to pay so there's no money to spare for petrol and bus fares. But we won't give up our freedom and we'll not let anybody forget what we did, what we showed them. The men were really shocked sometimes seeing how strong we were, but they used to sing, "We're proud of you, we're proud of you". We won't let them forget these times.

Women Involved in the Group: Tina Warboys, Elaina Stirland, Margaret Wood, Maggie Ryer, Pam Oldfield, Catherine Cooke, Jean Boddice, Rita Allen, Mrs. Wood, Kay Paling, Jane Marsden, Mary Sargison, Connie Birkin, Vickie Flude, Marlene Paling, Doris McGarry, Kim Cornell, Jill Montgomery.

WARSOP ACTION GROUP

At the beginning of April, 1984, a group of four friends, Bill Davies a striking miner, Val Davies and Bill and Elsie Cantrell, started to pack parcels with £30 from the union Hardship Fund. It was soon apparent that this team of workers and this income would be totally inadequate so as other women heard about the group, they joined and began to extend the activities. This was at first resented by one member of the original foursome and, looking back, the women realise how they were restricted. Vicky explains:

Because we weren't sure of ourselves at that stage and didn't really know what we were doing, we didn't stand up to this pressure. It stopped us from doing much more. We wanted to control things ourselves but we couldn't. Looking back we should have insisted.

I am sending £5. P.O.
I saw JACK TAYLORS LETTER
IN TRIBUNE. KEEP UP. THE GOOD
WORK. GOOD LUCK.
REDUNDANT MINER
AWINNIG COLL. 1969.
O.A.P. 76 YRS.
[illegible]

I am very pleased to send you a cheque for £5 on behalf of my wife and I. We are both senior citizens, and have been active in the working class movement the latter part of our life time.

I was working in the South Wales coalfield during the 1926 lock-out of the miners, and therefore appreciate the struggle that is going on to day. It is like living that period all over again. The Notts coalfield was then as it is now, the weakest link in our struggle to defeat the government and the Coal Board over pit closures.

The struggle of the miners is in the interest of the whole of the working class.

Best wishes to all in the front line

London S.W.8 May 1984

There were miners living in the immediate vicinity on strike from Welbeck, Warsop Main, Shirebrook, Whitwell, Thoresby, Sherwood, Blackwell Cable Workshops and Area Transport at South Normanton. Irene talks about how important it was to get premises:

> We wanted to start a kitchen but we never managed it. We got the Labour Hall for five days then that went down to three. There were problems all the time, some Labour councillors wanted us out, some supported us, some tried to be on both sides. We should have demanded facilities for cooking, we needed them. We listened to somebody else's advice and we're sorry now. We should have done what we wanted to do. It would have given us a kitchen for meals and somewhere to meet and talk. It would have been a social centre for us, like some of the other groups had. It's our greatest regret that we didn't go for what we wanted, that we let somebody put us off.

It was only possible, therefore, to provide food parcels. A typical parcel from Warsop consisted of:

4 lbs. sugar	2 tins of tomatoes	1 tin of spaghetti
1 packet of cheese	4 cans of beans	4 tins of soup
2½ dozen eggs	1 loaf of bread	soap
tinned or fresh meat	cereal	toothpaste

A local Greek fish shop owner gave wonderful support, sending potatoes and fish cakes every week. Warsop was twinned with Hull Dockers and Hull University and support also came from Huddersfield Labour Party. By the time the group was functioning properly, contacts had been made by other groups and there was little left for Warsop. Some women went to Hull to make the personal links which were so important in twinning and some went to Liverpool with Liz Hollis, collecting in pubs for the central funds.

> It was awful, asking for money from folks who were as badly off as us. But they gave us what they could. It was worse in Nottingham when we went to the stall outside Marks. We got a lot of abuse. At the time when Gadafi was linked with Arthur, people sneered at us and asked for Gadafi's phone number. We felt as if we were

> begging. Of course, we were begging, because we had to. The government would have liked us to be starved, wouldn't they?

The Warsop women are proud of the fact that they were the first to obtain housing benefit for single men on strike. This meant a lump sum of back payments for men who had hitherto received nothing in state benefits. Work by Danny Phillips and others at the Ollerton Welfare Rights Centre brought information on such claims and made a great difference to the position of single men.

By October the number of women left had been drastically reduced from about twenty four to six, so few were able to go out collecting or speaking. The men on strike were also far fewer and they were working round the clock providing logs, sawing, bagging and delivering.

At Christmas £10 was given for every child and each family had a turkey and other Christmas food. A Barnsley butcher sent specially prepared good meat, brought down by pickets, and occasionally food came from Derbyshire to help cater for men living in Nottinghamshire but on strike from Derbyshire pits.

One regret the women have is that they did not reach out to more wives of strikers.

> There were many who stayed at home all the time. They must have felt terrible, just being on their own, watching the news on TV and not doing anything. They didn't get the chance to learn the lessons we learned.

Mary Maddock feels that it would have been almost impossible to go out to the women in their homes:

> There was always too much to do, especially when our members went down. I was involved with Church Warsop group as well and we all had so many jobs. But I think we should have made a special effort.

Vicky, who was treasurer and secretary and was also the contact person for the Legal Centre at Ollerton, agrees:

But we were worked nearly to the end of our tether. I used to sit up late at night writing letters to supporters. It was important to reply to everyone, they deserved at least a letter when they were doing so much. I'd have enjoyed it if I hadn't been so dead tired and wanting to go to my bed.

In many ways the strike did us good, and the children as well, they learned to share and to go without things they didn't really need. I think we are all better people after the strike, poorer but better. We've picketed together, marched together, faced the police together, we needed a lot of courage to do that, courage we didn't know we had before the strike. You know, we're proud of ourselves and now we can do things for ourselves and by ourselves. We really know what is important now, what matters in life.

Women Involved in the Group: Vicky King, Irene Prichard, Mary Maddock, Joan Chiltern, Hazel Wain, Julie Hunter, Janet Dale, Debbie Pearson, Sheila Swallows, Joan Langley, Ann Mitchell, Val Davies, Barbara Ives.

By October only Vicky King, Irene Prichard, Val Davies, Mary Maddock and Ann Mitchell were left; women were expected to leave the group when their men went back to work.

WELBECK WOMEN'S ACTION GROUP

Welbeck was always in the news, both locally and nationally. There were so many problems with finding and keeping premises that there were always protests, lobbies and occupations and Welbeck became one of the best known groups. Throughout the strike, the women had their work constantly interrupted and disrupted by these difficulties.

Sandra Taylor, who worked as a collector for a firm, started to go round, house to house, with two union men from Warsop Main colliery, asking her customers for tins of food. Response was good and the food was distributed to families known to be on strike. Lynn Newburn called a meeting at her home and from that meeting the group was formed.

> We were the blind leading the blind. Nobody really knew what we were doing or what we could do. We started off going round Mansfield Woodhouse, knocking on doors, asking for food. We got about six carrier bags and plenty of abuse. We got something from Silver Saver and apologies from the big shops in Mansfield. By the end of April we knew there was a group organising in Mansfield and we got in touch with Ida Hackett. She told us to come to a meeting in the old library and we packed in, women from all over Nottinghamshire. They gave us £20 and we got a lot of ideas; we heard about picketing as well.

Sandra begins the saga of premises:

> When we had a big group of women we wanted a kitchen. The NCB let us have the Youth Club at first, but only to do food parcels and make tea. We thought we'd get away with soup as well, but after only two days somebody reported us and we were out. From then on, only the men could use it, and just for meetings. I went to see the committee and Jan Spencer went to Clowne to ask for financial help from the Yorkshire Area. I apologised for breaking the rules and promised to be good. But by this time a majority of men had gone back so they even took away the hall for meetings. Then they let us back in, but we couldn't do any cooking. We made a buffet, you should have seen it, it really looked good. The men thought it was fantastic.

At the same time, efforts were being made to acquire the Village Hall instead of the Youth Club. Social Services and some members of the Nottinghamshire County Council (Labour controlled) tried to intercede on the group's behalf, but the trustees were obstructive and employed all kinds of tactics to frustrate the women. For a period it was given, with restrictions, then taken away, then given again. In the meantime every family was receiving a food parcel of about £2.50 to £6 in value, depending on funds. During the summer holidays the school was available to prepare and serve meals for children but in September the group was back with the same problem.

> The Village Hall committee had got us out finally by saying we'd

had pickets sleeping in there. They said we'd gone over their heads to the County Council and Sandra was accused of having a key cut. They said anything to keep us out.

At last, tired and frustrated, they decided to stage a peaceful protest. It turned out to be the most brutal confrontation experienced by women in Nottinghamshire during the strike.

It was to be peaceful, women and children only, no men. It was to be like a procession, to see the pit manager and put our case for the use of the hall. We thought if we could explain our problems and he could see us with the kids he'd perhaps help us. There were some men standing at the side of the road, watching us. They were just standing there, some had kids with them, just watching. When we got to the place where the official pickets were, the police stopped us and told us to stand with the pickets, waiting for the scabs coming off shift. We told them we weren't pickets, we wanted to get to the pit to see the manager.

All the women who took part in the demonstration say that the scenes which followed are engraved on their memories for ever:

The police closed off all the roads. They had their helmets and chinstraps and truncheons. Then they began to attack the men on the side. They dragged away one who had a child with him; the little one was left crying. They tried to pull a child out of another's arms so that they could arrest him. The men weren't doing anything, just watching us. There were more men arrested in our village that day than on any other day of the strike.

★

All hell was let loose. folks were screaming, kids were crying, frightened to death. All unnecessary, we only wanted to see the pit manager. They wanted to show us they were not letting us; we had no rights, they could do what they liked.

After that there was no possibility of getting use of the hall, at least not by asking for it. Much earlier on, when problems over the hall had reached their worst, someone had lightly suggested an occupation.

Banner made by Dutch women supporters for Welbeck Women's Action Group; *Left to Right:* Janet Thompson, Margaret Graham, Ann Heaton

Women at Welbeck Campaigning for use of the village Hall

Everyone had heard of other groups—Clipstone, Blidworth and Rainworth—who had got premises by occupying them, and it seemed the only way. Jill Timmins, Jan Thompson, Carol Packman and Alyson Maggs now took the initiative in organising what they saw as their last desperate attempt to get somewhere to feed the men, women and children still out on strike.

> The plan was to take in sleeping bags and food for one day. We knew women would bring in what we needed after that. Somehow, everybody got to know what we'd done and loads of women came from other groups, congregating outside the hall, singing and shouting in solidarity. The reporters turned up and so did the police. We even had TV cameras and Welbeck Women's Action Group made the news. It got the trustees together and they finally agreed we could have the hall back, but only for women and children. We wouldn't agree, we were in and we were stopping until we got what we wanted.

In the meantime there was a lobby of women at Nottinghamshire County Council in Nottingham over the matter of school clothing allowances. The opportunity was taken to bring up the matter of Welbeck. A suggestion that the women should move to Warsop was resisted and Fred Riddell, Chairman of Education, used his influence to insist that the local school be offered instead of the Village Hall. There was much opposition from the headteacher and governors, but Riddell insisted and the women demanded a contract with all the conditions stated before they would agree.

At last they felt secure, despite attempts by some of the kitchen staff to intimidate them and obstruct their work. Numerous notices appeared all round the kitchen; pots, pans and salt pots mysteriously disappeared. Everything calculated to hamper and worry the women was done and there were constant complaints of misuse of the facilities. "We couldn't have left it cleaner, but they still complained, they hated us." Earlier in the year all the women in Nottinghamshire had been invited to a meeting in Ollerton where Brenda Greenwood had exhorted them to leave the kitchen sink to the care of the men and go out speaking. Sue Carlisle and Terry White in London had arranged bases for Welbeck men and so now Welbeck women decided to go and join in the collections.

> It wasn't long before we got into trouble with the Notts Forum. We went like lambs to the slaughter. We told the men we were desperate for money but they said that any money collected was to go to the Forum. We couldn't collect in their patch. But we didn't take any notice. Twelve of us started off on our own, to London, Monday to Friday, with the men or without them. Sometimes we did thirty meetings in a week, exhausted we were. You just wanted to sleep all weekend after a stint down London. NALGO in Newham adopted us and they said we could go anywhere in their meetings. It wasn't all good though. Sometimes people didn't show any interest, sometimes they were downright hostile, you never knew what you'd find. Sometimes the meetings had only a few people, that was miserable, sometimes they were packed, that was frightening. It was good at Universities and Colleges. Fancy us educating them. We were amazed how ignorant they were about the strike.

Sandra is still especially amazed at her ability to go to the European Parliament in Strasburg and speak about the strike, asking for support, "Who'd have thought, a year ago, I'd be doing that, not me!"

Supporters often seemed more dedicated than the women themselves were, wanting to go on and on when everyone was tired out. They never seemed to want to give in. So successful were the Welbeck women that they raised large sums of money and were able to do without support from central funds. Moreover, they were able to donate £6,195 to other groups in need and to send money to Wales and Yorkshire.

The strike taught them many lessons about compassion and comradeship as well as about guts and determination. They treasure their memories, despite the suffering and hard work. Their attitude to the miners who did not strike or who went back before the end is one of forgiveness. They explain that some were just not strong enough. Elsie says:

> We have to talk to them, we have to educate them as we have been educated or we shall get nowhere. We cannot hate for ever, but they must do their share and they don't show much sign of it yet.

We don't want to go on hating but we won't forget what the Nottinghamshire miners did by not coming out with us and winning the strike.

Women Involved in the Group:

Chairperson:	Lyn Newburn, then Sandra Taylor
Vice Chair:	Lynda Higginbottom, then Anne Heaton
Treasurer:	Jean Cockroft
Secretary:	Jill Timmins
Auditor:	Chris Gustard
Committee:	Carol Packman, Janet Thompson, Karen Meakin, Margaret Graham, Janet Spencer.

Other Women: Alyson Maggs, Marian Book, Pauline Hume, Elsie Henderson, Chris Dunlop, Cheila Burns, Chris Downer, Sue Lang, Edith Cameron, Patsie Collingwood, Jill Jarrett, Nora Batey, Lynn Davy, Iris Burton, Mavis Bowater, Jeanette Nicholson, Marjorie Brown, Ann and Lynda Taylor,Jean Cave, Hilary Gustard, Pat McPherson, Mary Baker, Heather Scott, Yvonne Brown, Marian Blower, Glenys Booth, Brenda Main, Audrey Davis, Mrs. Hopper.

WHEN THE END CAME

No woman seemed able to find words to describe adequately what she felt when the announcement came that the strike was over and the men had been told to return to work. From the moment the news came, they were thrown into a state of shock, numbed with sadness, even though they had known for weeks that events were moving rapidly towards the end. They had carried on with their work, trying to be cheerful and optimistic, pushing out of their minds the possibility that the National Coal Board would win and there would be a return to work without the agreement on pit closures which was central to the fight of the NUM. When the end finally came, the women were like characters in a drama on whom the final curtain had descended before they had finished playing their parts. The storm that had alternately raged and smouldered for a year was suddenly quenched. There was a terrible sense of loss. The women had not finished. They were impotent, frustrated, suddenly made purposeless by a decision totally outside their control.

The men went back to their pits. The centres which had been like home for so long closed down, one by one, if not immediately, then within a few days. Several more distributions of food parcels were made, to help tide over until the first wages were received. Money was still available for desperate cases from the hardship fund run by Pam Elliott and Henry Richardson. Food which was left in the communal freezers was shared or raffled with free raffle tickets, and winners took home their 'prizes', a packet of frozen sausages or a bag of frozen chips. The women cleared up and went home. Children continued to ask why they could no longer go to the kitchens, bewildered and disappointed that all the activity they had so much enjoyed was at an end. A pall of melancholy despair descended over the strikers, the women and children. There was a deep sadness that they had fought so hard and achieved so little. None of their demands had been won. The women particularly, knew underneath that they had achieved so much, if not their main intention. They had done many more things than they had set out to do when they had entered the arena of the dispute. But this was of little comfort in the days immediately after the strike. There were, inevitably, recriminations, particularly about the failure of the Trade Union and Labour movements to give the kind of support that the miners had called for. It was especially directed against Neil Kinnock and others in the Labour Party leadership whose support had been seen as derisory. They felt bitter and disillusioned. They questioned whether all the hardship had been to any purpose. They were sure that they could have won, given the full support of the men and women of their own coalfield.

And then, after the initial blow and the slough of despondency which accompanied it, there came attempts to see the positive side. There were brave statements that the battle had indeed been lost but that the war was not over. There was the cause to be fought of those miners who, as a result of the dispute, had been dismissed from their jobs or had been imprisoned. Funds were set up all over the country and speakers were needed to go out and make appeals for money. The women were desperately tired. They were angry when well-meaning people asked what they were going to do next, often retorting that they were going to have 'a bloody good rest'. The worries of managing domestic budgets cut to the minimum, of being cold and miserable in damp and desolate houses, of wondering who would go back next, of whether or not funds would dry up for the kitchen, of what the next attack from the courts, the police, the government and the Coal Board would be; all these had taken their toll on physical and mental well-being. The constant need to shop, cook, and prepare food parcels for strikers, to go out on speaking excursions all over the country and abroad, to attend meetings, rallies and demonstrations and to care for their children and dependants had made excessive demands on energies now depleted, worn down over the long, nerve-racking year. However, in spite of their exhaustion the women in Nottinghamshire rallied to the cause of the sacked amd imprisoned miners, continuing to go out to meetings to raise funds, continuing to set up stalls for collections at every event where people sympathetic to the cause would gather, continuing with the city stall in Nottingham.

Many pit based groups ceased to meet, though some women made their meetings into social events. They felt the need to stay together and enjoy the comradeship and pleasure of the year spent in such close proximity. The central group continued on a fortnightly basis, with much smaller attendances, but seeking to remain the focal point for a variety of activities.

Just as a kind of listless debility seemed to be affecting many of the former activists, there was suddenly a new demand for their energy and time. The leaders of the non-strikers in Nottinghamshire launched their campaign to vilify the leaders of the NUM, particularly Arthur Scargill, and to form a breakaway organisation. This attempt to inflict a serious and intensely damaging blow on the National Union was seen by the women in Nottinghamshire as the next target in their campaign, an opportunity to swing their full weight behind the fight-back to protect the NUM. They began again to show their resourcefulness and determination. The took up the cry, 'Keep Notts National', organised meetings, printed thousands of leaflets for distribution at pit heads and when they were stopped, took them out to the individual houses of the miners who had worked during the dispute.

Not all the women who where active in the strike continued to be involved in the two on-going campaigns. Some women, it is true, have gone back to the kitchen sink. But they have gone back physically, not mentally. They may well enjoy the daily routine of home, caring for dependants, making a pleasant and supportive environment for men and children. There is great value in caring for others in this way. What is of fundamental importance is that they may do very much the same things as they did before the strike, but they will never again think in the same way. Their new knowledge is indelibly impressed on their consciousness, their awareness and understanding heightened and extended, their self-confidence assured.

Whatever groups the women may join in the future—community associations, peace campaigns, political parties, environmental groups, clubs and federations—their experiences must be of advantage to other women campaigning for better conditions, for equal rights and opportunities, against exploitation, discrimination and prejudice wherever it exists. Whatever they wish to do, they need not look far, there is so much to be done. They are women who know that they are strong; they know their capabilities and they have much to offer.

APPENDIX I
MAJOR EVENTS IN THE STRIKE

(Events specific to Nottinghamshire in Italics)

1984

1 March — Closure of Cortonwood Colliery announced; Yorkshire miners come out on strike.

10 March — Conference of Branch officials agreed resolution re-affirming total confidence in Executive members.

15 March — David Jones, Yorkshire picket, killed at Ollerton, Notts.

Area Ballot: Do you support Strike Action to prevent Pit Closures and massive run-down of jobs? Result: 20,188 against, 7,285 in favour. Following the death of David Jones, Area Executive agrees to call all Notts Area members out on strike from 6.00. pm on Thursday, 15th. March, pending the result of the ballot.

18 March — Conference of branch officials to discuss ballot result. Decided on immediate return to work, with continuation of overtime ban.

19 March — Area Executive condems over-reaction of police operation in Notts, and in particular outside NUM Area offices on previous morning. Agreed the Executive should have a 24 hour duty rota at office.

21 March — Bolsover branch lobby of Executive following North Derbyshire Area decision to call members out on strike and subsequent mass picketing of Bolsover Pit.

At same meeting committee agree to advise branches that fundamental rights both to work and not to cross picket lines should be respected. Members refusing to cross picket lines advised

	to return home immediately.
23 March	Police seal off county of Nottinghamshire.
3 April	*Following the ASLEF decision to ask members to completely black all coal and coke movements, Notts Area Executive agreed to recommend the following resolution to a meeting of branch officials and committee members to be held on 5th April: "In view of the fact that other trade unions throughout Great Britain are instructing their members not to cross picket lines in support of the miners' fight for jobs, this Area must search its trade union principles regarding its policy on picket lines. We therefore recommend that branches set up their own picket lines and the Area Executive Committee instructs our members not to cross those picket lines. We reiterate our supports for a national ballot."*
5 April	*Above resolution rejected by meeting.*
11 April	NACODS vote for strike action, but vote does not reach the two thirds required.
17 April	*Special Council meeting to consider amendments to National Rule Book; 7 branches voted in favour, 24 against. Agreed that Area would pay for one coach from each branch to attend National Rally against pit closures to be held in Sheffield to coincide with Special Delegate Conference on 19th April.*
18 April	*Notts Area Hrdship Fund established; donations to come from other Notts Area Funds.*
19 April	NUM delegate conference votes against a national ballot.
25 April	*Scargill and Heathfield address conference of branch officials and committee members, stressing that the Resolution of the Special National Delegate Conference held on the 19th superceded Area decisions and was thus official NUM policy.*
26 April	*Area Executive agree to inform all branch officials that strike is official.*

12 May	Women Against Pit Closures hold 10,000 strong rally in Barnsley.
23 May	NUM leaders meet McGregor; talks abandoned same day.
5 June	*Reported to Area Council that National NUM had confirmed that strike action in Notts was offical according to National Rules.*
13 June	Further talks break down.
18 June	'Battle of Orgreave'—6,000 to 7,000 pickets; 93 arrested, 59 injured (including Arthur Scargill).
5/6 July	NUM and NCB meet for fresh talks.
8 July	*Control Centre at Berry Hill closed down.*
9 July	National Dock Strike called over use of scab labour to unload iron are at Immingham Docks.
9/10 July	*Sit-in at the Area Office, Berry Hill.*
18 July	NUM/NCB talks break down.
23 July	Dock Strike called off.
31 July	South Wales Area NUM fined £50,000 for contempt of court, assets seized.
10 August	NUM Special Conference calls on TUC for support.
13 August	*Area Executive agree that Harship Fund should be topped up from Area General Fund to ensure continuing weekly payments of £30 to each Women's Support Group.*
23 August	Second Dock Strike called over unloading of coal at Hunterston.
3 September	TUC conference votes to support NUM.
9 September	Talks start again between NUM and NCB.
15 September	Talks break down.
19 September	Second dock strike ends.
28 September	NACODS vote 82.5% in favour of strike.
1 October	Labour Party Conference shows solid support for NUM.
3 October	NACODS and NCB go to ACAS.
11 October	New NUM/NCB talks at ACAS.
15 October	Talks fail.
24 October	NACODS calls off threatened strike after NCB

	proposes changes in review procedure. NUM rejects that NCB proposals.
3 November	Price Waterhouse seize NUM assets on behalf of the sequestrator.
19 December	*National Union challenge Notts Area in high court over proposed rule changes. Mr. Justice Warner refuses injunction.*
20 December	*Nottinghamshire Area Council endorse rule changes and vote to loosen ties with National NUM.*
1985	
12 January	*Henry Richardson suspended from his position as General Secretary by special Council meeting.*
14 January	*Executive meeting suspends Henry Richardson from all duties at Area Office.*
24 January	NUM Executive meets NCB following meeting between Smith and Peter Heathfield.
29 January	Talks break down.
15 February	NUM and NACODS reject NCB's surrender document.
21 February	Special Delegate Conference rejects TUC's document.
25 February	*Overtime ban lifted by Notts Area.*
1 March	Durham, Lancashire, South Wales and COSA vote for a return to work without agreement.
3 March	Special Delegate Conference at TUC votes 98 to 91 for a return to work on March 5th.
5 March	Offical return to work. Many Yorkshire and South Wales miners refuse to cross picket lines put on by Kent miners.
10/11 March	All miners except those sacked or jailed return to work.
11 March	*Executive dismiss Henry Richardson from his positions as Area Official, Area General Secretary and Area Agent.*
18 March	*Special Executive meeting decides that Notts Area should not participate in the national ballot, nor should the special levy of 50p be implemented*

	in the Notts Area.
13 May	*Ballot of Notts membership asks whether members support Notts Area Executive Committee in opposing new national rule changes: 15,157 (73%) in favour, 5,631 against.*
22 May	*Ballot for new Area official to replace Henry Richardson—Neil Greatorex elected.*
1 July	*NUM Annual Conference in Sheffield. David Predergast and Roy Lynk dismissed by National Union of Mineworkers.*
3 July	*Nottinghamshire delegation walk out of Annual Conference.*
6 July	*Nottinghamshire Area dissociate from NUM.*
8 July	*Ray Chadburn, President of Nottinghamshire Area, dissociates himself from the breakaway. Ray Chadburn dismissed by Nottinghamshire Area.*
August	*Witch-hunt begins. Any member declaring support for the NUM is suspended.*
October	*Nottinghamshire and South Derbyshire Areas ballot on transfer of engagements and amalgamation to form a new organisation.*
6 December	*Union of Democratic Mineworkers receives registration from the Certification Officer.*
1986	
June	*4,800 members still remain in the NUM (Nottinghamshire Area) although National Coal Board refuse recognition.*

INDEX OF SUPPORT GROUPS

OTHER BOOKS FROM CANARY PRESS

STATE OF SIEGE
Jim Coulter, Susan Miller and Martin Walker

State of Siege documets and analyses the role of the police in the first six months of the 1984-85 miners' strike. Based on first hand interviews and research it is not only acknowledged as one of the most important books on the strike, but represents the first serious look at what are dangerous political developments with a much wider implication.
£3.50 240 pages November 1984

A TURN OF THE SCREW
Martin Walker

A Turn of the Screw is the story of an unfinished struggle. Following the year long 1984-85 miners' strike, a wave of repression swept over the coalfields. Less overt than that experienced during the dispute, this repression was economic, cultural and legal, harder to identlify than the violence handed out by the police on the picket lines. It was met with a series of short strikes as the miners continued to fight back. The strike ended with seven hundred men sacked, over sixty in prison and those back at work harassed by management and in some areas a breakaway union. A Turn of the Screw tells the story in the miners' own words, with analysis and short pieces of fiction. Most importantly the book describes the new consciousness developing among those who were victims of state violence, sackings and imprisonment.
£3.95 132 pages October 1985

SHIFTING HORIZONS
Lynn Beaton

Shifting Horizons describes the part played by two Nottinghamshire miners' wives in the 1984-85 strike. Both women were born and grew up in mining communities and during the strike they fought not just for their future but to preserve their history as well. But as they were to find out, this history had not prepared them for the struggle which was pressed upon them. Forced by events into a new awareness and identity, they found new roles as organisers and activists. Shifting Horizons charts a political journey of two working class women through Thatcher's Britain, two women who fought amongst thousands more for their communities and their dignity. A story of courage and fortitude, of resilience and learning, it is at times touching and filled with raw humour, at other times chilling and painful. It is the story of everywoman in struggle.

£5.65 265 pages October 1985

LET THEM EAT COAL

The Political Use of Social Security During the Miners' Strike

Lesley Sutcliffe and Brian Hill

Let Them Eat Coal explains the political and legislative background to the Thatcher government's policies on welfare benefits. It centres upon the treatment which striking miners in the 1984-85 strike received at the hands of the DHSS and is based upon information received from welfare rights workers throughout the country, together with first hand accounts and interviews with striking miners and their families. The booklet explains the political use of the benefits system by the Conservative government to break the National Union of Mineworkers and starve the strikers and their families back to work.

£1.95 55 pages February 1985

STRIKE BREAKING IN ESSEX

Moira Abdel-Rahim

Strike Breaking in Essex reports on the policing of the miners' picket lines at the Essex ports for the six months form the beginning of April to the end of September 1984. It also includes an examination of the Dartford Tunnel incident, roadblocks and other illegal police activity, the operation of the police 'mutual aid' scheme and an assessment of the cost of the operation. Based on interviews and first hand experience of the picket lines, it exposes not only the role of the police but the attitude of the Essex Police Authority and the view of the Chief Constables during a ruthless campaign to deny picketing rights, crush the miners and step up police powers.

£2.50 121 pages 1985

DEATH IN THE CITY
Melissa Ben and Ken Worpole

Death in the City is an examination of police-related deaths in London. Between 1970 and 1985 over 300 people died in London police stations or in other circumstances related to police activity. Towards the end of the 1970's the numbers rose dramatically. Based upon the personal experience of families and friends, and evidence given at inquests, the book documents some of these deaths and describes the struggle of individuals and communities who set out to find how they occured. What emerges is a catalogue of official indifference, neglect, institutional racism and sexism, human folly and legal maladministration. The book also identifies pointers to the incresing number of deaths in the growing powers of the police, their tactics and their organisation.
£3.95   122 pages   January 1986